With best Compliments

Ardsa [signature]

March 14, 2006

EMERGING BIOLOGICAL THREAT

NATO Science Series

A series presenting the results of scientific meetings supported under the NATO Science Programme.

The series is published by IOS Press and Springer Science and Business Media in conjunction with the NATO Public Diplomacy Division.

Sub-Series

I.	Life and Behavioural Sciences	IOS Press
II.	Mathematics, Physics and Chemistry	Springer Science and Business Media
III.	Computer and Systems Sciences	IOS Press
IV.	Earth and Environmental Sciences	Springer Science and Business Media
V.	Science and Technology Policy	IOS Press

The NATO Science Series continues the series of books published formerly as the NATO ASI Series.

The NATO Science Programme offers support for collaboration in civil science between scientists of countries of the Euro-Atlantic Partnership Council. The types of scientific meeting generally supported are "Advanced Study Institutes" and "Advanced Research Workshops", although other types of meeting are supported from time to time. The NATO Science Series collects together the results of these meetings. The meetings are co-organized by scientists from NATO countries and scientists from NATO's Partner countries – countries of the CIS and Central and Eastern Europe.

Advanced Study Institutes are high-level tutorial courses offering in-depth study of latest advances in a field.
Advanced Research Workshops are expert meetings aimed at critical assessment of a field, and identification of directions for future action.

As a consequence of the restructuring of the NATO Science Programme in 1999, the NATO Science Series has been re-organized and there are currently five sub-series as noted above. Please consult the following web sites for information on previous volumes published in the series, as well as details of earlier sub-series:

http://www.nato.int/science
http://www.springeronline.nl
http://www.iospress.nl
http://www.wtv-books.de/nato_pco.htm

Series I. Life and Behavioural Sciences – Vol. 370 ISSN: 1566-7693

Emerging Biological Threat

Edited by

George Berencsi

National Center for Epidemiology, Budapest, Hungary

Akbar S. Khan

US Army Edgewood Chemical Biological Center, USA

and

Jiři Haloužka

Institute of Vertebrate Biology, Academy of Sciences of the Czech Republic, Department of Medical Zoology, Valtice, Czech Republic

IOS Press

Amsterdam • Berlin • Oxford • Tokyo • Washington, DC

Published in cooperation with NATO Public Diplomacy Division

Proceedings of the NATO Advanced Research Workshop on Emerging Biological Threat
Budapest, Hungary
5–8 October 2003

ISBN 1-58603-555-X
Library of Congress Control Number: 2005930832

Publisher
IOS Press
Nieuwe Hemweg 6B
1013 BG Amsterdam
Netherlands
fax: +31 20 687 0019
e-mail: order@iospress.nl

Distributor in the UK and Ireland
IOS Press/Lavis Marketing
73 Lime Walk
Headington
Oxford OX3 7AD
England
fax: +44 1865 750079

Distributor in the USA and Canada
IOS Press, Inc.
4502 Rachael Manor Drive
Fairfax, VA 22032
USA
fax: +1 703 323 3668
e-mail: iosbooks@iospress.com

PRINTED IN THE NETHERLANDS

Emerging Biological Threat
G. Berencsi et al. (Eds.)
IOS Press, 2005

Opening Remarks

Ladies and Gentlemen,

It is my privilege to welcome you on behalf of the Head of the Health Service of the Hungarian Defense Forces, on behalf of Major General László Svéd, M.D.

The date 6th of October is a historical Anniversary in the Hungarian Republic. In 1849 the Army of the Russian Czar defeated the defense forces of the Hungarian Democratic Revolution raised against the Austrian Habsburg Empire, and on the 6th of October 13 Generals were sentenced to death and executed by the Austrian government in the city of ARAD, located since 1919 in Romania.

The title of this 2003 conference is “Emerging Biological Threat”. We believe that this advanced research workshop will be one of the best initiated by the NATO Scientific department, and organised by Co-Directors Akbar Khan and George Berencsi. The topic itself has Hungarian roots.

The first microtechnique in microbiology was developed by Dr. Takátsy in the early fifties. The microplates and spiral loop system used for many decades for serological diagnostic has Hungarian origins.

This Advanced Research Workshop has limited funding. The title is very progressive, since emerging infections are concerned, but top experts in this topic are available in the United States. The Sponsor decided not to prefer training courses held by US professionals for Hungarian and eligible countries. In spite of this, we believe that this meeting will be very advantageous for participants.

There are many reasons for optimism. American, European and Hungarian participants are represented in equal proportions. I have to mention, that in 2003 the first Hungarian writer Imre Kertesz, was awarded with the Nobel Prize. Natural sciences have to aim to be at the same level.

I am sure that the material of the Workshop will be of outstanding scientific quality. I should like to wish you success in your work, and please prepare your manuscripts in time to be included in the edited proceedings of this meeting.

Budapest, 6 October 2003

Colonel István Kopcsó, MD
Hungarian Defence Forces,
Institute of Health Protection

Contents

Opening Remarks v
Colonel István Kopcsó

Part 1. Biological Threat Caused by Emerging and Re-Emerging Infections

Interrelationships of Emerging Bacterial and Viral Infections in Hungary Before the Introduction of Active Immunisation 3
György Berencsi, Peter Gyarmati and Zoltan Kis

Response to the HIV-Related Challenges in Lithuania 10
Saulius Caplinskas

Emerging Viral Hepatitis B and C in Estonia 20
Ljudmilla Priimägi, Valentina Tefanova and Tatjana Tallo

Epidemiology of Influenza A(H1N1) as One of Emerging-Reemerging Diseases 26
A.N. Slepushkin, E.I. Burtseva, L.N. Vlassova and V.T. Ivanova

West Nile and Other Emerging-Reemerging Viruses in Russia 33
D.K. Lvov, A.M. Butenko, V.L. Gromashevsky, M.Yu. Shchelkanov, A.I. Kovtunov, K.B. Yashkulov, A.G. Prilipov, R. Kinney, V.A. Aristova, A.F. Dzharkenov, E.I. Samokhvalov, H.M. Savage, I.V. Galkina, P.G. Deryabin, B.Ts. Bushkieva, D.J. Gubler, L.N. Kulikova, S.K. Alkhovsky, T.M. Moskvina, L.V. Zlobina, G.K. Sadykova, A.G. Shatalov, D.N. Lvov, V.E. Usachev and A.G. Voronina

Natural Foci of Classical and Emerging Viral Zoonoses in Hungary 43
Emőke Ferenczi, Gábor Rácz, Gábor Faludi, Alíz Czeglédi, Ilona Mezey and György Berencsi

Coronavirus Causing Severe Acute Respiratory Syndrome (SARS), a New Highly Infectious Emerging Threat 50
Younes Ali Saleh, Péter Gyarmati, Zoltán Kis and György Berencsi

Antiviral Drugs for Treatment of Herpes B Virus Infections 58
Maria M. Medveczky, George E. Wright, Richard E. Eberle and Peter G. Medveczky

Tick-Borne Bacterial Diseases in Hungary 60
Andras Lakos

Emerging or Re-Emerging Virus Diseases: Genetic Variation, Immune Failure or Human Mistake 62
J. Rajčáni and J. Pastorek

Part 2. Development of the Diagnostic Technology to Cope with Emerging Threat More Efficiently

Real-Time PCR for Detection of Parenterally Transmissible Viruses 77
Mario Poljak, Boštjan J. Kocjan, Katja Seme and Dunja Z. Babič

Diagnostics Development Research Within the U.S. Biological Defense Research Programme 84
George V. Ludwig

New Molecular Targets for Control of *Yersinia Pestis* Infection 93
Wieslaw Swietnicki, Kamal U. Saikh, Teri Kissner, Beverly K. Dyas, Afroz Sultana and Robert G. Ulrich

Enhanced Concentration and Extraction of Bacillus Anthracis DNA from Whole Blood 101
Matt Ewert

Biological Toxins and Super-Antigens as an Emerging Biological Threat 104
Akbar S. Khan and James Valdes

Application of Genomics and Proteomics for Detection Assay Development for Biological Agents of Mass Destruction 109
Vito G. DelVecchio

Development of Reagent Kits for Detection of Lethal Toxins 117
Gábor Faludi, István Jankovics, Ildikó Visontai, Júlia Sarkadi and Gyöngyi Zelenka

Part 3. Bioterrorism as an Emerging and Reemerging Biological Threat

Strategies for the Detection of Unknown Biological Materials 125
Peter J. Stopa and Jeff Morgan

Clinical Aspects of Bioterrorism (Anthrax, Plague and Smallpox) 136
László Rókusz

Clinical Virologists and the Smallpox-Threat: A Comparison Between Germany and the UK 146
Joachim J. Bugert

Genetic Attributes for Tracking Unconventional Microorganisms 147
Abdu F. Azad

An Attempt to Characterize Some Soil and Health Relevant Bacteria by FT-IR Spectroscopy 148
Zdenek Filip, Susanne Herrmann and Jaromir Kubat

Advanced Medical Technologies Against Bioterrorism 157
M. Garstang and E. Busch-Petersen

Science for Peace Means International Cooperation to Control and Prevent Emerging Natural or Bio-Terror Virus Epidemics with Attention to the HIV-1/AIDS Pandemic 159
Yechiel Becker

Contributors 174

Author Index 181

PART 1

BIOLOGICAL THREAT CAUSED BY EMERGING AND RE-EMERGING INFECTIONS

Emerging Biological Threat
G. Berencsi et al. (Eds.)
IOS Press, 2005

Interrelationships of Emerging Bacterial and Viral Infections in Hungary Before the Introduction of Active Immunisation

György BERENCSI, MD, PhD, Peter GYARMATI and Zoltan KIS
Division of Virology, "B. Johan" National Center for Epidemiology, Budapest Hungary

Respiratory infections have been reported in Hungary since 1932 on monthly basis to the National Institute of Public Health. These infections were complicated influenza like illnesses, measles, varicella-zoster, scarlet fever, epidemic meningitis, pertussis and diphtheria. The reexamination of these epidemics indicated, that the influenza epidemics were associated with the significant reduction of the number of clinical illnesses caused by other respiratory virus infections. The spread of enteroviruses during summers contributed probably to the decrease of infections caused by the above viruses, too. Epidemics of respiratory illnesses of bacterial etiology were not influenced by influenza epidemics, with the exception of *N. meningitidis* epidemics, which had been positively influenced by influenza and probably by other viruses (i.e. Respiratory Syncytial Virus). Combinations of microorganisms may significantly increase or reduce virulence or increase pathogenicity of each other. Combination of microorganisms may result in or interfere with emerging infections at the level of microbial physiology.

1. Introduction

The therapeutical application of the growing number of industrial recombinant interferon preparations for the treatment of chronic hepatitis patients and other illnesses has revealed, that the majority of the patients treated with several million units of recombinant (alpha) interferon weekly were suffering from influenza like clinical symptoms.

It has been published many years ago, that enteroviruses possess significant interfering capibility. Coxsackievirus B3 has prevented the progress of poliovirus type 1 epidemic in 1958 in Hungary [1–4]. The spread of enterovírus 71 could be prevented by the vaccination with oral poliovirus vaccine [5], and a nation-wide epidemic of echovírus 11' has been stopped by the vaccination of about 5 per cent of the children's cohorts in 1989 in Hungary with monovalent OPV type 1 [6]. It has been also proven, that between 1962 and 1972 the yearly vaccination of cohorts between 3 to 39 months of age have reduced the circulation of non-polio enteroviruses by 80 to 50%. It has been also detected, that interferon-sensitive and interferon-resistant echovirus 7, 11 and 14 serotypes were circulating among the vaccinees time to time in the country [*Kapusinszky B., Szebeni B., et al. unpublished results*].

It has been suggested by several authors [3], that the symptoms of influenza like illneses (ILI) are caused by the overproduction of interferons and other mediators induced by the infection. The aim of this work was to detect possible interactions of ILI and other viral and bacerial illnesses in ages, before the introduction of active immunization in Hungary.

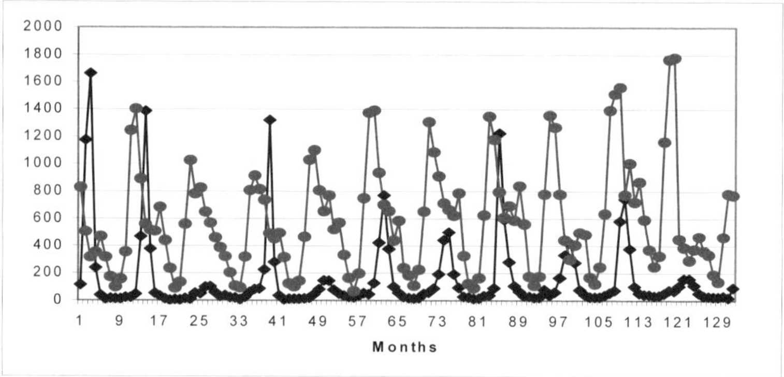

Figure 1. Varicella-zoster virus epidemics (grey curve) and the association of the monthly number of reported illnesses with that of complicated influenza like infections (dark curve). *Abscisse:* number of months elapsed after the 1st of January, 1932. *Ordinate:* number of reported illnesses registered every month.

2. Methods

The National Institute of Public Health has been founded in 1927 in Budapest, supported by The Rockefeller Foundation. The "paper-based" epidemiological reporting system has been immediately introduced for all infectious diseases, which could be reported on the basis of the clinical data by family doctors.

Only the respiratory illnesses have been compiled in these reports. The reports have been submitted every month to the National Institute, and registered. Respiratory infections reported were as follows

- Complicated influenza-like illnesses;
- Varicella-zoster (VZV) infections;
- Measles infections;
- Epidemic meningitis infections;
- Scarlet fevers;
- Diphtheria illnesses;
- Pertussis infections.

The original reports got lost during the Second World War, therefore, the data could be introduced into the Microsoft Word Excel format only recently. The data have been taken from the published tables of the National Institute of Public Health compiled by Béla Johan and Joseph Tomcsik [2].

3. Results

3.1. Effect of Influenza Epidemics to the Spread of Other Respiratory Viral Illnesses

The interaction of complicated influenza like illnesses and varicella-zoster virus (VZV) epidemics between 1932 and 1942 are plotted in the 1st figure. Each influenza epidemic resulted in sharp decrease of the number of VZV infections diagnosed. It is surprising, however, that in certain years a second decrease of reported clinical VZV diseases could be observed (in the 49th, 89th and 113th months in Fig. 1). The decrease in number of VZV illnesses was independent of the severty of influenza epidemic (i.e. the degree of interferon

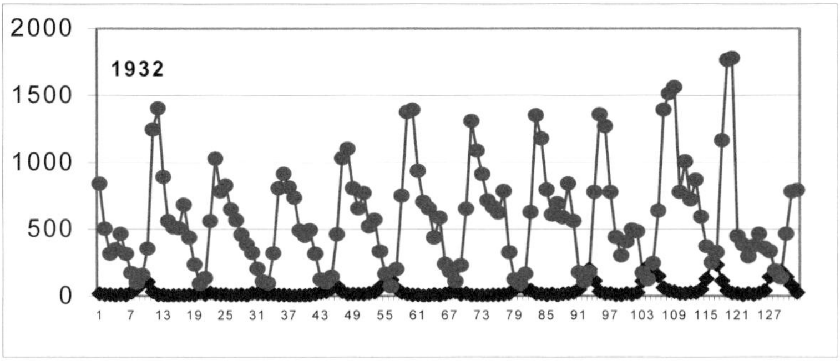

Figure 2. Interaction of Varicella-Zoster diseases (grey curve) with epidemics of poliomyelitis (dark curve) between 1932 and 1943 in Hungary. *Abscisse:* number of months elapsed after the 1st of January, 1932. *Ordinate:* number of reported illnesses registered every month.

induction was probably independent of the virulence of the influenza variant). Significant decrease of the reported VZV infections were registered in years when the number of complicated influenza like illnesses was very low (see month No. 49 in Fig. 1).

3.2. Interrelationship of Poliomyelitis Epidemics and the Number of Varicella-Zoster Illnesses

On the basis of the published data enterovirus epidemics exert significant interfering activity to simultaneously circulating polioviruses [3,5,6]. Timely association of poliomyelitis epidemics to the registered VZV illnesses are summarised in Fig. 2. In some cases the maximum of poliomyelitis peaks were ovelapping with the minimum number of VZV illnesses (57th, 78th and 127th months). In other years, however, the minimum of VZV infections preceeded the maximum of paralytic poliomyelites (8th, 43th and 92th month in Fig. 2). The second sharp decrease of registered clinical VZV illnesses observed in certain years have not be found to overlap with paralytic poliomyelitis (i.e. enterovirus) epidemics. One has to hypothesize, that an other viral agent had been responsible for the abrupt and significant induction of interference comparable to that induced by influenza epidemics.

During the period examined only paralytic poliomyelitis cases could be registered. One has to be aware, however, that the circulation of polioviruses were preceded or followed by the circulation of other enteroviruses every summer. The interrelationship of VZV illnesses and poliomyelitis epidemics are shown in Fig. 2.

It might be of interest, that the minimum of VZV and also measles infections was overlapping with the maximum circulation of enteroviruses. The relationships of measles and influenza epidemics is shown in Fig. 3.

Accumulation of the number of patients suffering from *influenza complicata* was found to be followed by the significant reduction of the number of acute measle infections similar to the data observed in the case of clinical VZV illnesses. The pattern of measles, however, also indicated that in certain years a second virus caused induction of interference affecting considerable proportions of the younger cohorts reducing transiently the number of acute clinical measle infections. The peaks of influenza complicata illnesses appeare to be very low, since the number of reported measles cases were regularly higher, than those of illnesses of VZV etiology.

One can put forward several suggestions on the basis of the data shown in Figs 1 to 3.

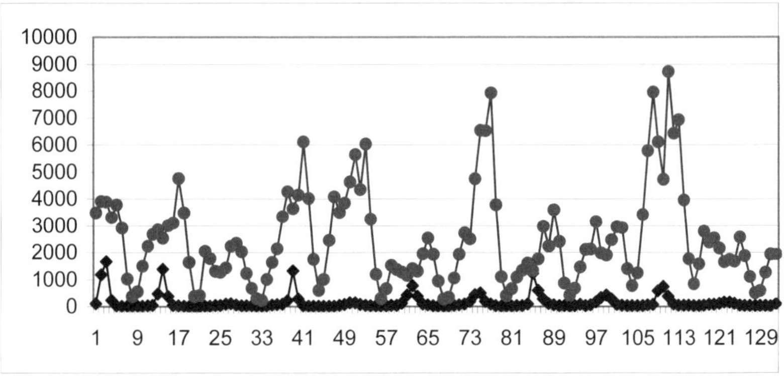

Figure 3. Influence of epidemics of influenza like illnesses (squares) to the dynamics of measles epidemics (circles) between 1932 and 1943 in Hungary plotted on the basis of the reports collected by *Bela Johan* and *Joseph Tomcsik*. *Abscisse:* number of months elapsed after the 1st of January, 1932. *Ordinate:* number of reported illnesses registered every month.

1) Both clinical VZV and measles illnesses were found to decrease every year in association with the maxima of the of influenza epidemics. On the basis of the contemporary information available, one may suggest, that the acute overproduction of interferon and other mediators might be the cause of this observation.
2) The virulence of influenza virus variants circulating was not associated with the rate of reduction of other viruses transmitted by the respiratory route, since the reduction could be seen also in years, when the number of *influenza complicata* cases reported were low (months 28, 50 and 120–121 in both Figs 1 and 3).
3) The observations seem to indicate that in four seasons during the years of this survey a second interferon-inducing etiologic agent (probably virus) has caused epidemics in Hungary affecting mainly children. The curves of both reported viral diseases measles and VZV revealed sharp decreases without any overlapping with the influenza epidemics or circulation of enteroviruses (months 25, 49–50, 57–58, 97 and 105–113 in both Figs 1 to 3). Epidemics preceding the seasons of influenza like illnesses could have been viral gastroenteritis epidemics.
4) The interaction of enteroviruses and respiratory pathogens may suggest, that the abrupt decrease of the later during the summer might be the consequence of the interfering ability of enteroviruses. No such interrelationship could be seen with respiratory infections of bacterial etiology.

4. Interactions of Bacterial and Viral Infections

In connection with the epidemic of Spanish flu in 1918 and 1919, *Haemophilus influenzae* was isolated from the majority of patients. It could be shown only decades later, that the host cells killed by the virus provided the bacteria with growth factors required for multiplication (i.e. heme and nicotinic acid-adenine dinucleotide). Following the sessation of virus replication the overproduced bacteria die because of the lack of growth factors. Release of endotoxins from the lysed bacteria might result in complications or relapses due to destruction of new cells by endotoxins or due to the delayed regeneration of mucus membranes.

In recent years, influenza like illnesses (ILI) are frequently caused by procaryotic pathogens causing interstitial pneuminae (i.e. *Mycoplasma pneumoniae* and *Chlamydia pneumoniae)* sometimes alone, but frequently in combination with viruses. Severe complications and lethal outcome can be prevented by the application of 2nd or 3rd generation of macrolide type antibiotics.

Enhancig cooperation of staphylococci and influenza viruses have been discovered also by Rudolph Rott and coworkers in Giessen, Germany several decades ago [7].

In the United Kingdom, epidemic meningitis occurred at the end of the thirthies, too. Military camps had been affected at the first place, since many young soldiers had been accommodated in large dormitories (*Dr. T.M. Reid,* personal communication). One of the local epidemiologists has recognized that the epidemic could be stopped when the distance between heads of two soldiers was more than three feets from each other during the night. With this discovery, they have decided to change the position of every second person in the dormitory. The positions of soldiers were head to feet in the neighbouring beds. The epidemic stopped within one week in the camp. The first explanation of the empirical observation has been provided by Raza *et al.*, 1993, 1994 and El Alhamer *et al.* 1996 & 1999 [8–11]. They presented evidence, that *N. meningitidis* was unable to penetrate tissue culture cells without the helper effect of respiratory syntitial virus (RSV). Upon RSV infection a new membrane protein appeared on the cytoplasmic membrane of the cells enabling *N. meningitidis* to adsorb and penetrate the infected cells only.

5. Epidemiological Association of Influenza Epidemics and Accumulation of Clinical *Neisseria Meningitidis* Diseases

The first epidemic of *N. meningitidis* which could be analysed by epidemiologists was caused by the serogoup B between 1939 to 1942. A complete 3300 clinical epidemic meningitis cases were registered during the three years from a 10 million population. Never again occurred more then 150 clinical diseases per year in Hungary, although serogroups A, C, Y and Z had been introduced into the country between 1996 to 2001. The monthly number of serogroup B epidemic is shown in Fig. 4.

Clinical illnesses caused by *N. meningitidis* were accumulated every year several weeks after the accumulation of severe influenza illnesses. This phenomenon cannot be explained only by epidemiological rules. The new serogroup entered the country at the end of the influenza season in 1939 and the number of clinical meningococcal diseases has decreased paralell with the number of complicated influenza patients (Fig. 4). A minor peak of ILI was seen on the fall of 1940 without any influence on epidemic bacterial meningitis. The influenza season, however, in 1940, 1941 and 1942 has been associated with the reemergence of the disease.

In 1941 three overlapping epidemic meningitis peaks could be observed. Currently it has been detected that the circulation of RSV happens around December before the influenza season in Hungary [3,12–14]. The corresponding shoulder of the ILI peak in 1942 can be seen in Fig. 4.

One may suggest on the basis of the data collected in the era of diagnostic virology, that RSV could be responsible for the outbreak of *N. meningitidis* before the influenza epidemics.

The biochemical explanation of the enhancing effect of influenza viruses to N. meningitidis became available in the 70th and 80th. Vaccines were developed and manufactured in order to immunize children against epidemic meningitis.

The only *N. meningitidis* serogroup against which no vaccine could be manufactured was serogroup B. The reason is, that the chemical composition of the capsule of serogroup

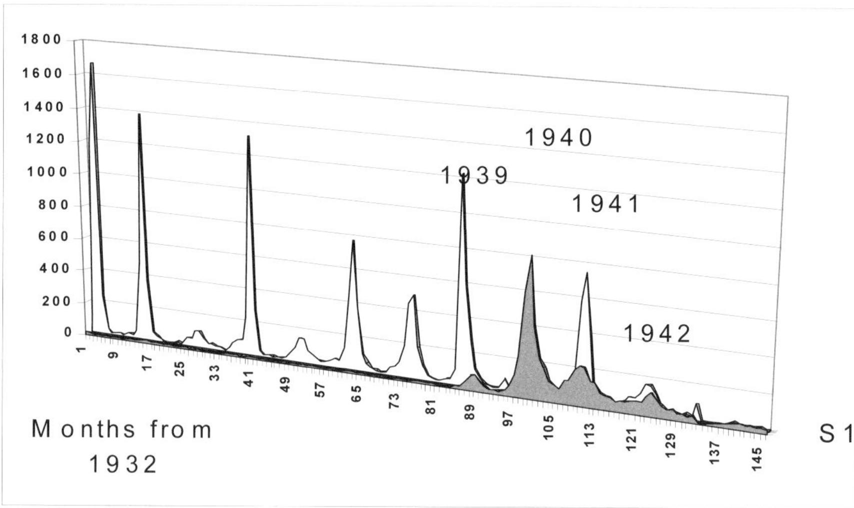

Figure 4. Number of severe influenza patients per month between 1932 and 1943 in Hungary (white peaks) and number of clinical epidemic meningitis patients registered between 1939 and 1942 in Hungary (grey peaks).

B is composed exclusively of neuraminic acid [15]. Neuraminic acid (sialic acid) is, however, present on the surface glycoproteins of the majority of the human cells. The immunization containing only capsule polysaccharides of *N. meningitidis serogroup* B (i.e. neuraminic acid oligomeres) cannot be successful.

Neuraminic acids, however, are receptors for all influenza vírus types and participate in the adsorbtion of RSV, certain parainfluenza viruses and even in the case of G2 coronaviruses.

According to the epidemiological observations our hypothesis is, that influenza viruses can be absorbed by *N. meningitidis*, and the bacterium-virus complex may be adsorbed by human cells. These complexes may be taken up by the human cells, but this is never the case of individual bacteria, since the capsule is able to prevent even phagocytosis by macrophages due to the identical sugar components of human glycoproteins and bacterial capsules.

Neisseria meningitidis serogroups other than B have been imported in Hungary only after 1996. Fortunately never such large number of clinical illnesses occurred, probably because of the following reasons:

1) Capsules of serogroups A, C, Y and Z are composed of neuraminic acid and other sugars [15]. These capsules are probably less active in the absorption of influenza viruses.
2) The currently circulating influenza types and subtypes caused only smaller epidemics affecting only 3 to 5% of the population.
3) These risk groups of the population have been vaccinated every year in the last decade in Hungary. These vaccinations reduced probably further the size of the influenza epidemics in the country thus decreasing the risk of the penetration of meningococci into the blood streem or cerebrospinal fluid of the seronegative persons.

About 600 clinical epidemic meningitis cases have been registered by the Department of Epidemiology during the years concerned. These complex situation may provide probably the explanation why the invasions of 4 new serogroups of meningococci resulted in 5 times less clinical diseases, than that between 1939 to 1942.

6. Conclusions

The interactions of bacteria, RSV, influenza and other respiratory viruses indicate, that combinations of microorganisms may significantly increase or reduce virulence or increase pathogenicity of each other. Combination of microorganisms may result in emerging infections at the level of microbial physiology. The genetic interactions represent a different level of the modification of pathogenicity and virulence.

References

[1] Dömök, István: From influenza to AIDS, results, analyses and problems (in Hungarian) Egészségtudomány 41, 100–116, 194–210 (1997).
[2] Year Books of the National Institute of Public Health between 1932 to 1943, Eds. Bela Johan and Joseph Tomcsik (in Hungarian).
[3] Berencsi Gy., N. Szomor K., Gyarmati P.: Influenza epidemics in a changing world (in Hungarian). Magyar Orvos 10, 46–47, 2002.
[4] Dömök I., Molnár E.: Bornholm epidemic in 1958 in Hungary (in Hungary). Orv. Hetil. 101: 1306–1313 (1960).
[5] Nagy G., Takatsy S., Kukan E., Mihaly I., Domok I.: Virological diagnosis of enterovirus type 71 infections: experiences gained during an epidemic of acute CNS diseases in Hungary in 1978. Arch Virol. 1982, 71(3):217–27.
[6] El-Sageyer M.M., Szendrõi, A., Hütter, E., Új, M. *et al.*: Characterization of an echovirus type 11` (prime) epidemic strain causing haemorrhagic syndrome in newborn babies in Hungary. Acta Virol. 42, 157–166, 1998.
[7] Tashiro M., Klenk H.D., Rott R. Inhibitory effect of a protease inhibitor, leupeptin, on the development of influenza pneumonia, mediated by concomitant bacteria. J Gen Virol. 1987, 68:2039–41.
[8] Raza M.W., El Ahmer O.R., Ogilvie M.M., Blackwell C.C., Saadi A.T., Elton R.A., Weir D.M.: Infection with respiratory syncytial virus enhances expression of native receptors for non-pilate Neisseria meningitidis on HEp-2 cells. FEMS Immunol Med Microbiol. 1999, 23:115–124.
[9] El Ahmer O.R., Raza M.W., Ogilvie M.M., Blackwell C.C., Weir D.M., Elton R.A.: The effect of respiratory virus infection on expression of cell surface antigens associated with binding of potentially pathogenic bacteria. Adv Exp Med Biol. 1996, 408:169–177.
[10] Raza M.W., Blackwell C.C., Ogilvie M.M., Saadi A.T., Stewart J., Elton R.A., Weir D.M.: Evidence for the role of glycoprotein G of respiratory syncytial virus in binding of Neisseria meningitidis to HEp-2 cells. FEMS Immunol Med Microbiol. 1994, 10:25–30.
[11] Raza M.W., Ogilvie M.M., Blackwell C.C., Stewart J., Elton R.A., Weir D.M.: Effect of respiratory syncytial virus infection on binding of Neisseria meningitidis and Haemophilus influenzae type b to a human epithelial cell line(HEp-2). Epidemiol Infect. 1993, 110:339–347.
[12] Hoekstra R.E., Herrmann E.C. Jr., O'Connell E.J.: Virus infections in children. Clinical comparison of overlapping outbreaks of influenza A2-Hong Kong-68 and respiratory syncytial virus infections. Am J Dis Child. 1970, 120:14–16.
[13] Nicholson K.G.: Impact of influenza and respiratory syncytial virus on mortality in England and Wales from January 1975 to December 1990. Epidemiol. Infectol. (1990) 116:51–63.
[14] Fleming D.M., Cross K.W.: Respiratory syncytial virus or influenza? The Lancet (2001) 342: 1507–1510.
[15] René Germanier (Ed.): Bacterial vaccines. Academic Press, New York, 1984.
[16] Diet N.H., Libikova H.: Selective resistance to togaviral superinfection in mice with tolerant lymphocytic choriomeningitis virus infection. Acta Virol. (1979), 23:385–392.
[17] Diet N.H., Libikova H., Rajcani J.: Suppression of an arenavirus by a togavirus in experimental acute double infection. Acta Virol. (1978), 22:391–400.

Emerging Biological Threat
G. Berencsi et al. (Eds.)
IOS Press, 2005

Response to the HIV-Related Challenges in Lithuania

Saulius CAPLINSKAS, MD, PhD
Lithuanian AIDS Center

The global HIV/AIDS epidemic killed more than 3 million people in 2003, and an estimated 5 million acquired the human immunodeficiency virus (HIV) – bringing to 40 million the number of people living with the virus around the world. Every day about 14.000 persons in the world contract HIV [1]. This report concentrates to the preventive efforts introduced and run in Lithuania.

Eastern Europe

The data of UNAIDS shows that there have been steady increases in the number of people living with HIV/AIDS, as well as in the numbers of AIDS death. The Eastern Europe continues to experience expanding epidemics, with the number of people living with HIV/AIDS growing year by year.

The AIDS epidemics in Eastern Europe shows no signs of abating. Some 230,000 people were infected with HIV in 2003, bringing the total number of people living with the virus to 1.5 million. AIDS claimed an estimated 30,000 lives in the past year.

Worst-effected are the Russian Federation, Ukraine, and the Baltic States (Estonia, Latvia, and Lithuania). Driving these epidemics is widespread risky behaviour – injecting rug use and unsafe sex – among young people [1].

Extraordinarily large numbers of young people regularly or intermittently engage in injecting drug use, and this is reflected in increasing HIV prevalence among injecting drug users throughout the former Soviet Union. Condom use is generally low among young people, including those at highest risk of HIV transmission in Eastern Europe. In Estonia and Latvia, it has been estimated that up to 1% of the adult population injects drugs. Overall, up to 25% of injecting drug users are estimated to be under 20 years of age across Eastern Europe. And the use of unclean equipment, often through sharing of drug injecting equipment, remains the norm. Young people predominate in this region among reported HIV cases. The young men bear the epidemic's brunt [1].

Another striking pattern is now evident. Women account for an increasing share of newly diagnosed HIV infections – 33% in 2002, compared to 24% a year earlier. One consequence is a sharp rise in mother-to-child transmission of the virus. There patters are most evident in regions where the epidemic took hold several years ago, such as Kaliningrad. Because most injecting drug users are young and sexually active, a significant share of new infections is occurring through sexual transmission (often when injecting drug users or their HIV-infected partners engage in unsafe sex) [1].

Baltic States

Although overall numbers of infections remain low, HIV spread continues at an alarming pace in the Baltic States. At 2,300 in 2002, the total number of HIV diagnoses in Latvia has risen five-fold since 1999. Just four years ago, Estonia reported 12 new HIV cases; in 2002, 899 people were newly diagnosed with the virus. Lithuania is on a similar path. Here, the 72 new HIV cases detected in 2001 increased more than five-fold in 2002. Lithuania appears to be facing two distinct epidemics – one affecting mainly injecting drug users in regions adjacent to Kaliningrad (Russia), and the other spreading among men who have sex with men in Vilnius [1].

Current data based only on people who are tested for HIV, and not all potentially affected groups of people are being tested.

Western Europe

The total number of people living with HIV continues to rise in high-income countries. The number of annual AIDS death has continued to slow in Western Europe, due to widespread availability of antiretroviral treatment.

The resurgence of other sexually transmitted infections in Western Europe points to a revival of high-risk sexual behaviour – especially among young people, including men who have sex with men. In Western European countries that report HIV cases, heterosexual intercourse may now be the most common mode of HIV transmission.

In Western Europe, just over 10% of newly diagnosed HIV cases in 2002 were caused by injecting drug use [1].

Lithuania

In 2002 397 new cases of HIV were registered in Lithuania, most of them at the Alytus penitentiary. The majority of the HIV-positive persons are males who contracted HIV through intravenous drugs.

As many as 735 HIV positive persons have been registered in the country before January 1, 2003. A year ago the number of those infected was 338, the number was twice smaller than this year. Doctors maintain, doubling of the number of HIV carriers was influenced by HIV outbreak at the Alytus penitentiary where 299 new HIV cases were identified. Most of the HIV carriers – as many as 90 percent – contracted HIV through intravenous drugs.

Specialists say, implementation of drug prevention measures is insufficient. In the West European countries the number of HIV cases among intravenous drug users has been falling, but there are more persons who contract HIV through sexual intercourse.

Last year in Lithuania 10 persons contracted HIV through sexual intercourse, out of them 3 are women.

In 2002, 9 new cases of AIDS were registered in Lithuania, in 2001 the number was 10. AIDS has been diagnosed to 55 persons. Out of 735 HIV-positive or those suffering from AIDS 41 are already dead (23 of them died of AIDS and the remaining passed away for unrelated causes). In 2002 4 persons died of AIDS and in 2001 the number was twice smaller.

Up to 1st November 2003, 97 new cases of HIV infections were reported, totaling to 832 HIV infected people. The predominant HIV transmission mode remains the same as in previous years: 73 people contracted HIV through contaminated injecting equipment, majority of them are males.

Situation Background

The presence of modernized infrastructures for both HIV prevention and drug treatment in Lithuania allowed for decisive action, which has been credited for quenching the initial outbreak of HIV infection among IDUs in the country and keeping the incidence low compared to the neighbours. Until September 2002 Lithuania's reported HIV infection rates were among the lowest in Europe [2].

Following recommendations of United Nations and World Health Organization, the National AIDS Prevention Programmes, attributable to priority national public health programmes, have been developed and approved by the Government.

In 1996, the AIDS Prevention Programme was included into the list of state compulsory health programmes by the order of the Government of the Republic of Lithuania, therefore it has the status of the State programme. AIDS issues are included into the Health Programme approved in 1998 by the Seimas (Parliament) of the Republic of Lithuania. It is envisaged to stop the increase in the cases of HIV and AIDS by the year 2010 [2].

In accordance with current epidemiological situation, the Action Plan of the National AIDS Prevention Programme has been corrected every year, and best practices of other countries put into practice.

The first National HIV/AIDS Prevention Programme was developed and implemented in 1990–1992. The later programmes in 1992–1993, 1994–1996, 1996–1998 and 1999–2001 followed an aim to reduce the spread of HIV in Lithuania and to implement other not only medical, but also social objectives.

The National AIDS Prevention and Control Programme 2003–2008 was declared a priority national health programme by Parliament of the Republic of Lithuania Resolution No VIII-833 of 2 July 1998, "regarding adoption of the Lithuanian Health Programme" (Official Gazette., 1998, No 64-1842), therefore this programme is an important part of the Lithuanian Health Programme [2].

Priorities of this programme have been set taking account of the rapidly changing epidemiological situation in Lithuania and neighbouring countries, HIV transmission trends, experience of health care and other professionals and latest scientific achievements, assuring continuity of the key measures of the previous programmes.

The programme was developed taking account of the WHO and UNAIDS AIDS strategy and policy programme documents.

Strategic Plan, Prevention, Human Rights, Care and Support

Lithuania has developed multisectoral strategies to combat HIV/AIDS including health, education, labour and science sectors. HIV/AIDS issues are integrated into the National Health Programme. A functional HIV/AIDS body assisting in the coordination of civil society organizations is the Lithuanian AIDS Center. Strategy that addresses HIV/AIDS issues among Lithuanian national uniformed services, including armed forces and civil defence forces is considered in the National AIDS Prevention Programme. However Lithuania still does not have neither functional national multisectorial HIV/AIDS management/coordination body, nor functional national HIV/AIDS body that promotes interaction among government, the private sector and civil society, nor has evaluated the impact of HIV/AIDS on its socioeconomic status for planning purposes.

Lithuania has a general policy to promote information, education and communication (IEC) on HIV/AIDS realized by the Lithuanian AIDS Center. The law on endorsement of policy promoting reproductive and sexual health education for young people was passed in 2002. Lithuania has also strategy that promotes IEC and other health interventions for groups with high or increasing rates of HIV infection, which include IDUs, sex workers, youth and prison inmates as well as strategy that promotes IEC and other health interventions for cross-border migrants. Executing agency is the Lithuanian AIDS Center. There is also a strategy and policy adopted to expand access, including among vulnerable groups (IDUs, prison inmates, sex workers, MSM), to essential preventative commodities, which include condoms, sterile needles, HIV tests. Policy and strategy to reduce mother-to-child HIV transmission is included in the National HIV/AIDS Prevention Programme for 2003–2008 that is still considered by the Government.

Lithuania does not has special laws and regulations that protect against discrimination of people living with HIV/AIDS. The country policy is not to discriminate this group of general population with special laws and regulations. General laws and regulations focusing on schooling, housing, employment, etc. are also applied to protect people living with HIV/AIDS. There is also a valid law on confidentiality protection. However there are laws and regulations that protect against discrimination groups of people identified as being especially vulnerable to HIV/AIDS, i.e. MSM, IDUs, prison inmates. Lithuania also has a policy to ensure equal access for men and women to prevention and care, which is put into practice by the Gender Equal Rights Commission at the Lithuanian Parliament. The policy ensuring that HIV/AIDS research protocols involving human subjects are reviewed and approved by an ethics committee exists.

Issues of promoting comprehensive HIV/AIDS care and support with emphasis on vulnerable groups are included into the National HIV/AIDS Prevention Strategy (under development), however up to now those have not been practically applied. Lithuania has a clear policy and strategy to ensure and improve access to HIV/AIDS related medicines with emphasis on the following vulnerable groups: IDUs, sex workers, MSM, youth, prison inmates. Commodities ensured are antiretrovirals, drugs for the mother-to-child prevention in pregnant HIV positive women, palliative care. So far, Lithuania does not have a policy or strategy to address the additional needs of orphans and other vulnerable children due to the relatively law HIV prevalence in the country.

Lithuanian AIDS Center

Systematic HIV/AIDS prevention in Lithuania was introduced in 1989 after the establishment of the Lithuanian AIDS Center.

The Lithuanian AIDS Center is in charge of HIV/AIDS prevention and control, epidemiological surveillance and data evaluation, public education, laboratory and confirmatory HIV testing, health care of people with HIV and AIDS, and is involved in psychological social rehabilitation of drug addicts. Priority activities of the Center relate to HIV positive drug addicts.

HIV prevention in the most vulnerable population groups has been stimulated and supported by international organizations and their representations. In 1997, a programme to support the target groups including sex workers, drug addicts, seafarers, etc. was launched with the aim to provide targeted information and improve accessibility of required medical and social services. In 1998, the Lithuanian AIDS Center with direct support from Vilnius Municipality opened the so-called "low threshold" site "Demetra" to assure medical, social and psychological services to these people.

The AIDS Center includes the following departments: Laboratory, Outpatient Department along with Anonymous Testing Site, Women's Health Site and Syringes Exchange Room, Public Education Department and Dependence Diseases Department. The Center also operates an AIDS Hot Line and a Hot Line for drug use affected people. Working fields of the Center are: care and treatment of people with HIV and AIDS, public education, laboratory testing (the laboratory deals as a reference one, the only in Lithuania performing confirmatory tests), anonymous testing and counseling, epidemiology surveillance and data evaluation, social rehabilitation of drug users, AIDS advocacy, policy and strategy, work with vulnerable groups.

With an active advocacy and lobbying of the AIDS Center, changes in laws and regulations, such as infectious disease control, homosexuality and testing policy, have been adopted since 1991, increasing the opportunities for preventive activities and making the testing more efficient. Approaches to STD control, highlighting the early diagnosis and comprehensive treatment, including counselling, have been introduced in some institutions thanks to the AIDS Center's efforts. Condom programming has been adopted as the only strategy to prevent HIV transmission, comprehensive approaches to drug use problem, including the harm reduction strategies, such as syringes exchange programmes, have been initiated. The AIDS Center beneficially prioritises primary prevention and exploits existing structures of health, social security, education and force to establish co-operation with local bodies. As the driving force, it operates as coordinating body with representatives of various sectors, ministries, municipalities and NGOs.

Along with establishment of the AIDS Center, basics for observation, treatment and follow-up of HIV positive persons and those with AIDS on out-patient basis have been set. All HIV positive persons, after diagnosis confirmation by the Laboratory of the AIDS Center, are being directed to the Dispensary Department to constantly observe the progress of the disease and prescribe proper treatment regimen. Since 1996 the Department also operates the Hot Line.

The staff is actively involved into the education of medical workers on various HIV/AIDS/STD issues, including prevention including training courses and lectures for a large medical audience. Lecturing already took more than 11000 hours, audience totaling to 13500 people.

In 1991 an anonymous room has been established to make the services more acceptable and available for the target groups who do not attend the polyclinics, to get in contact with this part of population, to better learn the epidemiological situation, behaviour and habits of people at greatest risk to get infected with HIV/STD. The personnel includes gynecologist and dermatovenerologist. Anonymous services were of great help to start targeted education of sex workers.

Principles of the Anonymous Room operation: voluntary, anonymous testing and strict confidential treatment. Services are charged.

After several years of the work on the streets trying to get in contact with sex workers, the Women's Health Site was opened with support of Vilnius Municipality in April 1998. Contacts settled through the anonymous room were succesfully used for staffing (former drug users, pimps, sex workers). The Site has followed an aim to reduce the harm in injecting drug users and street prostitutes. It has also involved into outreach activities. The Site also runs the Hot Line. Activities are based on the following principles: outreach work, volunteerism, free of charge services and confidentiality, peer education.

Up to now about 2000 active drug users and 500 sex workers have visited the Site. Presently, drug users are served in separate place – Syringes Exchange Site (Vytenio St. 37).

In 1991 the Lithuanian AIDS Center has announced two slogans: Lithuanian without Drugs and Harm Reduction for Drug Users, which marked initiation of actions targeted at drug users. Those included syringes exchange, support for the people who decide to quit using drugs, and establishment of drug users social rehabilitation unit (1992). This Department, presently called Dependence Diseases Department, enrolls men and women voluntarily applying for help after detoxification. First patients entered the Community in January 1993. Treatment duration is 12–14 months. Each patient is treated according to individual therapy and rehabilitation programmes, individual and group psychotherapy methods are applied.

Operation of the Dependence Diseases Department is based on the principles of family life. The community is situated in a cosy wooden house. During the leisure time the patients can listen to the music, watch video, read books, go in for sports, and work in the joiner's shop. Outings to the city are often organised.

The personnel consist of a psychotherapist, psychologists, who help the drug-users to regain physical and spiritual powers and find their place in life.

The principles followed: voluntary, conscious attendance and absolute confidence. Targeted attempts are made to further involve the former drug users into the peer education.

In 1993–2002 totally 107 drug users were treated in the Department, of them 65% have quit using drugs for more than one year. 55% completed successfully the full treatment course. Today all of them have jobs, many have changed their living place and got married. Many ex-patients still keep in touch with the community, some of them work in the field of education, perform preventive work among school children and youth. Those, who break up the course, received a short-time psychotherapy, which allowed further 11 patients quit using drugs.

Dependence Diseases Department is Open to Society

On Fridays the Department welcomes all people who seek for help or are interested in the Department's life.

A drop in day-care center for adolescents and young people was opened at the Dependence Diseases Department of the Lithuanian AIDS Center in 2000. Social counseling is being mixed with a variety of activities to motivate the children to change their life style. Contacts are established with parents and other relatives of the children. The day-care center offers the children possibilities to get involved in various courses, such as to learn foreign languages, to deal with computers, to see theatre, to get engaged in sport, etc. The visitors of the day center also attend social rehabilitation courses.

The laboratory of the Lithuanian AIDS Center is one of the most modern serology laboratories in Lithuania. A variety of tests on HIV/STD, opportunistic infections, virus infections, etc. are available in the Lithuanian AIDS Center's Reference laboratory. The most modern diagnostic methods have been first in Lithuania introduced and since applied in this laboratory. The Laboratory is the accredited member of WHO European Region Poliomyelitis Laboratory Network. Some of the tests performed:

1. Confirmation WB and LIA reactions for HIV, HTLV, HCV and syphilis
2. Diagnostic amplification tests for HIV, chlamydias, gonorrhea and TB
3. Diagnostic virological tests, e.g., HIV extraction and establishment of subtypes
4. Hematology tests, research of a variety of biochemical, cell and humoral immunity
5. Molecular both quantitative and qualitative HIV, HCV, HBV tests for diagnostic and treatment monitoring purposes
6. Laboratory care for HIV positive people
7. HCV genotyping
8. Laboratory diagnostics of sexually transmitted and blood-borne infections

9. Laboratory diagnostics of tick-borne diseases by serological and virological methods
10. Virological and serological diagnostics of influenza and other respiratory infections
11. Participation in the international projects ESEN2, SPREAD, etc.
12. Virological surveillance of poliomyelitis and other enteroviruses in Lithuania
13. SARS laboratory diagnostics by procedures of molecular virology.

Public Education Department

The main objective of this Department is to provide accurate information on HIV/AIDS, STI transmission modes and drug use prevention means to various population groups (medical workers, teachers, journalists, school children, students, etc.). The goal of great importance is to help people to adopt the rules of safer behaviour and change simultaneously the public opinion towards people living with HIV and AIDS, hence fighting stigma and discrimination. In this respect, a variety of mass events have been organized, e.g., annually (till 1997) organised "Condom feasts": shows with participation of popular musicians, groups, showmen. Mass events of this kind were aimed at young people just to brake the silence about all the safer sex-related issues.

The same aim has been followed by publication of targeted periodicals: AIDS Bulletin and AIDS Express Information – for medical workers, since 1989 newspaper "AIDS Chronicle" – for youth. AIDS Chronicle was further substituted by the bimonthly journals "Protect your Health" published in Lithuanian and "Mezdu Nami" (Between Us), published in Russian language – since 1994. The two latter publications became independent and have been published with some financial contribution from the AIDS Center [2].

Need for Dual Prevention

Consumption of illegal drugs is a well-grounded concern of the society [3]. Over the past decade, drug dependency had increased in Lithuania. In 1991, 15.3 per 100 000 inhabitants were drug dependent, and in 2001 – 101.2 per same number of inhabitants. Chemical consistence of used substances cannot be forecasted, thus there exists a real danger to poison oneself after one single trial [4]. The number of high school students having tried illegal drugs increases; in 1995 they made 3%, and in 1998 – as many as 22.7% [5]. Using of unsterile syringes and needles is threatening, as a result of it many drug users get infected with hepatitis B and C and HIV. Moreover, HIV is spread by unsafe sexual practises of IDU, as they often engage in prostitution in order to earn money for purchasing drugs and thus spread HIV to the population that does not consume drugs [6,7]. 40% of all AIDS cases in European region are caused by intravenous drug consumption [8]. HIV infected drug users make 80% of all HIV infected population in Lithuania [9]. At present experts all over the world including Lithuania do not limit themselves to just diagnosing AIDS cases and HIV prevalence surveillance. Hundreds of millions $ are spent for diminishing the virus spread. Presently, the need for dual prevention – concerted actions in the field of HIV and drug use prevention and control is obvious:

1) By its nature "dual prevention" requires collaboration between various services in health, social and other sectors.
2) Need to remain focused on the goal of preventing HIV/STI/Drugs and flexible to reach that through any means.
3) New alliances, new flexible approaches, new ways of thinking needed.

4) Demedicalisation of HIV/AIDS and related problems.
5) Accessible and acceptable health services for not insured high risk groups.

One of significant strategic objectives of HIV/AIDS prevention is drug use reduction and harm reduction associated with drug use. The greatest efforts are directed towards development of HIV prevention programmes of broad profile and their urgent implementation in order to increase motivation of safe behaviour among certain subpopulations [10–13].

Increasing HIV prevalence in target groups in Lithuania is an indicator, showing that more attention should be given to the development of preventive programmes that would accentuate service availability and acceptability to representatives of target populations. Relatively low HIV prevalence among the country's population means that either HIV preventive measures applied up till now were successful, or the virus has not reached the critical mass in the population yet. Monitoring of risky behaviour and targeted interventions till the virus has not reached the critical point provides with an excellent opportunity to plan and implement preventive strategies and thus diminish the risk of HIV spread. As indicated in the guidelines of WHO [14], the key principals of successful HIV prevention among IDU are: active contacts with IDU; availability and acceptability of medical and social support; information and health education; supplying with sterile injecting equipment and disinfectants; substitute treatment; use of condoms, diminishing the number of sexual partners and other.

Number of the drug users rehabilitation communities in Lithuania has increased since 2001. Lithuanian government started pay attention to drug use problem. Social Ministry of Lithuania initiated a project for running drug users rehabilitation programmes. The Lithuanian AIDS Center created successful model of rehabilitation community for drug addicts and shares experience with interested parties. The Center also initiated development of the National Drug Policy Plan.

The Plan includes wide spectrum of monitoring and evaluation programmes:

1. Monitoring drugs epidemics.
2. Monitoring STI epidemics.
3. Monitoring virus hepatitis B and C epidemics.
4. Monitoring HIV epidemics.
5. Monitoring interventions:
6. behavioural changes,
7. biological markers,
8. other
9. Comparing countries.

Survey of IDUs

The risk of individuals to get infected with HIV varies. Most often risky behaviour is concentrated in subpopulations, i.e. target groups. That is why, in order to determine the level of risky behaviour in IDU subpopulation, the survey of IDU visiting LAC Women Health Site with harm reduction unit "Demetra" on a constant basis was performed. Data on behavioural patterns of target groups reveal the main risk factors of contracting and passing HIV infection, and points out the specific behaviour that should be influenced by adequately used targeted interventions. Surveys of this type help to form more comprehensive preventive strategies and support managers of health and social sector in their decision-making. Coordinated measures implemented in target groups destroy HIV transmission chain in separate subpopulations and in general population as well. Without evidence-based information on HIV risky behaviour, as well as monitoring and analysis, neither health specialists nor specialists of other related sectors can plan complex interventions and obtain positive results while diminishing HIV spread among the population [15].

The targeted work with IDU performed in LAC Women Health Site with harm reduction unit "Demetra" was included into UNAIDS "Best Practice" bulletin in 1999 [16].

Specialists of LAC Social Disease Consultation Center "Demetra" questioned 142 intravenous drug users with the help of an adapted questionnaire for persons with high risk of contracting HIV infection, developed by Family Health International Impact [17]. The choice of IDU population for the behavioural study was determined by the fact that more than a half of HIV diagnosed persons in Lithuania belong to this societal layer.

The survey has showed the following. The majority of intravenous drug consuming subpopulation are young people with high school (secondary) education, whose average age is 22.9±6.5 years. 62.7% of the respondents indicated that they started consuming narcotic substances at school. The most popular substances are intravenously used opiates and opiates with sedatives. Parents learnt very late about their children's drug consuming habits, i.e. averagely after 2.7±2.7 years. Sharing of used injection equipment is rather widely spread in the surveyed subpopulation of IDU, showing that they do not have strong convictions about safe behaviour. Nevertheless, positive changes in the surveyed subpopulation were observed (percent of IDU who did not inject with needles and syringes used by other people increased twice, percent of IDU using new needles and syringes increased three times, as well as percent of those who did not give away, lend or sell used needles and syringes to other people). This shows that the interventions applied in "Demetra" that are targeted to IDU are acceptable and effective. Three-four casual sexual partners in last 30 days and rare condom use reflect rather high risk that HIV infection in IDU subpopulation may be transmitted sexually as well, and may be spread to subpopulations that do not consume drugs. While constantly performing health education and increasing the motivation to adopt safe behaviour, we managed to acquire positive changes in sexual behaviour of the surveyed population.

The survey results reflect that interventions provided by LAC Social Disease Consultation Center "Demetra" (exchange of needles and syringes, health education, communication and consultations) as well as available and acceptable opportunity to receive medical and social services confirm the hypothesis, that the behaviour of IDU may be influenced while directing it towards less risky tendencies, thus diminishing the risk of HIV spread to general population.

Conclusions

It is time to entirely rethink the former medical standards and those of social life. The problem of HIV/AIDS cannot be solved in the frame of traditional health and welfare system. HIV problem must be solved by people who understand that today we need not a formal work but personal participation, and that it is necessary to take over experience of people and successful organisations in time. The problem of post-soviet and post-totalitarian thinking is "to do what is necessary to do", without fear of ideological and other contradictions.

Two basic principles should be followed:

1. Fully accept high risk groups as equals,
2. Openness about sexuality and drugs.

Prevention of HIV among drug users, prisoners and other particularly vulnerable groups is still the most important and effective strategy. Prevention must be comprehensive and include support, care and option for antiretroviral therapy. Improvement of life skills of young people is the primary target area for health promotion activities. Changes in the society must be implemented to make improvement of life skills of the young possible. A system of treatment and rehabilitation has to be created so as to satisfy various needs of a person with dependency disease. We should seek a balance between the services of outreach, harm reduction, treatment, rehabilitation and social reintegration.

Currently Lithuania has certain favourable opportunities and perspectives to stay a low HIV prevalence country, to provide care and support to HIV infected and affected people. However, much effort has to be given to make the solution of HIV/AIDS related problems integrated, systematic, and effective. Neither targeted preventive interventions nor wider health promotion activities can be successful without "life skills of the society", i.e. non-discrimination, mobilisation of the entire society and sufficient funding. To be effective, the public health response to HIV will have create the social. Legal and ethical environment that is conductive to HIV prevention, care and support.

References

[1] AIDS Epidemic Update. 2003, Dec / UNAIDS/WHO – Geneva, 2003. – P. 3–14.

[2] National HIV/AIDS Prevention and Control Programme 2003–2008. – Vilnius: Lithuanian AIDS Center, 2003. – P. 3–9.

[3] Behavioural peculiarities of intravenous drug users and their vulnerability to HIV infection / Saulius Caplinskas, Irma Mittiene // Acta medica Lituanica. – 2001, suppl. 6, p. 42–47.

[4] Lietuvos sveikatos statistika 2003 = Lithuanian health statistics 2003 / Lietuvos sveikatos informacijos centras, 2003 = Lithuanian health information center 2003.

[5] The European School Survey Project on Alcohol and Other Drugs: the 1995 and 1999 ESPAD report. – Stockholm, 1997–1999.

[6] Prevalence of human immunodeficiency virus infection and behaviors associated with its transmission among parenteral drug users selected on the street / A. Rodes, M. Vall, J. Casabona, M. Nuez, N. Rabella, L. Mitrani // Med. Clin. (Barc.). – 1998, vol. 111, no. 10, p. 372–377.

[7] Prevalence of human immunodeficiency virus and risk behaviours among opioid users seen in an emergency room / P.K. Gonzalez, A. Domingo-Salvany, R. Hartnoll // Gac Sanit, 1999, vol. 13, no. 1, p. 7–15.

[8] Health 21 / WHO Regional Office for Europe. – Copenhagen, 2000. – P. 90.

[9] Data of Lithuanian AIDS Center 2003. – http://www.aids.lt.

[10] Guidelines for second generation HIV surveillance / WHO/UNAIDS. – Geneva, 2000.

[11] Project neighborhoods in action: an HIV-related intervention project targeting drug abusers in Washington / J.A. Hoffman, H. Klein, H. Crosby, D.C. Clark, J. Urban // Health. – 1999, vol. 76, no. 4, p. 419–434.

[12] Physician prescribing of sterile injection equipment to prevent HIV infection: time for action / S. Burris, P. Lurie, D. Abrahamson, J.D. Rich // Ann. Intern. Med. – 2000, vol. 133, p. 218–226.

[13] Structural interventions to reduce HIV transmission among injecting drug users / D.C. Des Jarlais // AIDS. – 2000, vol. 14, suppl. 1, p. 41–46.

[14] Principles for Preventing HIV Infection among Drug Users / WHO Regional Office for Europe. – Copenhagen, 1998.

[15] Drug injecting and HIV infection: global dimensions and local responses / WHO. – [Geneva], 1998. – (Social aspects of AIDS).

[16] Summary booklet of Best practice: Demetra Social Disease Consultation Center / UNAIDS Best Practice collection 1999, June. – p. 109. – http://www.unaids.org.

[17] Behavioral Surveillance Surveys. Guidelines for repeated behavioral surveys in populations at risk of HIV: implementing AIDS Prevention and Care Project / J. Amon, T. Brown, J. Hogle, J. MacNeil, R. Magnaui, S. Mills, E. Pisani, T. Reble, T. Saidel, C.K. Sow // Family Health International Impact. – 2000.

Emerging Biological Threat
G. Berencsi et al. (Eds.)
IOS Press, 2005

Emerging Viral Hepatitis B and C in Estonia

Ljudmilla PRIIMÄGI, Valentina TEFANOVA and Tatjana TALLO
National Institute for Health Development, Virology Department, Tallinn, Estonia

High prevalence of drug use during last years resulted in the significant increase in morbidity with hepatitis B and C and concentrated HIV epidemic in Estonia. Even the onset of the hepatitis A virus epidemics were shown to be associated with intravenous drug use. Characteristics of these emerging viral infections are summarised below.

Estonia is a country in Eastern Europe on the shore of the Baltic Sea, which was a Soviet republic until it regained its independence in 1991. Estonia is the smallest of the three Baltic countries with a population of about 1.4 million. More than two-thirds of the population live in urban areas in North, including the capital Tallinn, and North-Eastern part of the country, close to the border with the Russian Federation. Basic statistical data show that the health of Estonian population has been worsened since 1990; the death rates attributable to cardiovascular diseases, accidents, and poisonings have all risen. Life expectancy was lowest in 1994, being for men 61.1 and for women 73.1 years, the former having declined from 66.5 years and the latter from 74.9 years in 1988 [9].

Though some signs of improvement have been noted in recent years, the incidence of some infectious diseases is still high. Hepatitis B virus (HBV) and hepatitis C virus (HCV) infections with parenteral mode of transmission have spread more widely in Estonia during last decade. The incidence of acute hepatitis B and C increased 2.9 and 4.7 times in 2001, respectively, compared to 1995, being 32.8 and 22.4 per 100 000 population (Fig. 1, data of Health Protection Inspectorate of Estonia).

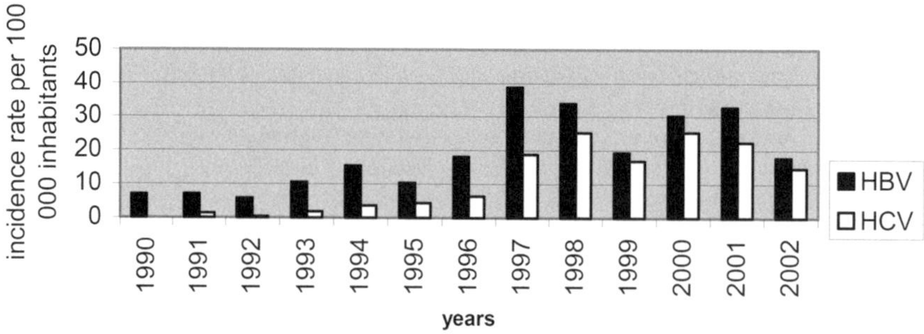

Figure 1. Incidence of viral hepatitis B and C per 100 000 inhabitants in Estonia (1990–2002).

Table 1. Share of viral hepatitis B and C cases in Northern and North-Eastern parts of Estonia, 1997–1999.

		HBV			HCV		
		1997	1998	1999	1997	1998	1999
Northern- and North-Eastern part of Estonia	abs.	533	452	234	269	353	177
	%	**94,3**	**91,3**	**83,6**	**98,9**	**96,2**	**72,5**
Estonia (total)	abs.	565	495	280	272	367	244

Data of Health Protection Inspectorate, Estonia.

Table 2. Reported incidence of acute HB and HC per 100 000 population by age in 1996, 2000 and 2001, in Estonia.

Age group	Incidence rate of HB per 100 000 population			Incidence rate of HC per 100 000 population		
	1996	2000	2001	1996	2000	2001
0–14	3.0	2.0	6.6	1.3	3.2	5.4
15–19	71.4	175.0	182.4	25.7	162.4	115.5
20–29	53.0	88.8	87.1	19.2	69.2	58.1
30–39	18.7	22.0	19.5	7.5	18.8	17.3
40–59	5.6	8.2	8.4	1.1	5.4	6.8
60 >	3.6	4.5	3.8	0.7	1.4	2.1

Data of Health Protection Inspectorate, Estonia.

The rapid increase of morbidity with hepatitis B and C started from 1996–1997, mainly in Northern and North-Eastern part of Estonia. The number of registered cases of viral hepatitis B and C there made up about 90% of all registered cases in country (Table 1).

The registration of chronic hepatitis, which may be a source of hidden infection, has also been started in 1998 in Estonia. During 1998–2002 there was found steady increase in number of registered chronic hepatitis: 1998 – 6.8%, 1999 – 12.2%, 2000 – 17.3%, 2001 – 23.2%, 2002 – 34.5% (data of Health Protection Inspectorate).

Overall prevalence of HBV and HCV infections was the highest in age 15–19 and 20–29 and among men (Table 2).

Injection drug use is the main mode of transmission of these infections among youth, accounting for more than half of newly acquired infections (Tables 3 and 4). According to our data [8] the seroprevalence of HBV and HCV infections among IDUs (visitors of Anonymous consulting rooms) were 72.3% and 91.5%, respectively. Prevalence was highest among imprisoned HIV-positive IDUs – 91.1% and 97.5% in 2002 respectively. Most of these persons are chronically infected, but not aware of their liver disease, as being not clinically ill.

The spread of viral hepatitis preceded the concentrated epidemic of HIV infection in 2000–2002 among IDUs in North-Eastern and Northern Estonia.

During 1988–1999 the incidence of detected HIV cases was low and stable in Estonia, ranging from 0.1 to 0.7. HIV was spreading equally through homosexual (38.2%) and heterosexual (33.7) contacts and there were no cases among pregnant women and infants.

Table 3. Risk factors, associated with reported cases of acute hepatitis B in 1997–2001 (%).

Risk factors	Percent of patients with acute hepatitis				
	1997	1998	1999	2000	2001
Blood or plasma transfusion	0.9	0.2	0.4	0.5	0.9
Medical manipulation	3.4	2.8	7.1	1.8	2.9
Working with blood	0	0.4	0.7	0	0
Drug addiction	46.0	33.9	50.4	50.1	57.2
Sexual contacts	4.2	4.0	3.9	4.1	5.1
Tattoos	–	–	–	–	0.7

Data of Health Protection Inspectorate, Estonia.

Table 4. Risk factors, associated with reported cases of acute hepatitis C in 1997–2001 (%).

Risk factors	Percent of patients with acute hepatitis				
	1997	1998	1999	2000	2001
Blood or plasma transfusion	0	0	0	0	0.3
Medical manipulation	0.7	1.6	2.5	1.6	1.0
Working with blood	0	0	0.4	0.8	0.3
Drug addiction	51.8	62.7	54.9	48.8	52.9
Sexual contacts	3.7	1.1	2.1	3.0	5.9
Tattoos	–	–	–	–	2.6

Data of Health Protection Inspectorate, Estonia.

In late 2000 there was a rapid increase in incidence of new HIV cases in IDU subpopulation in North-Eastern part of Estonia which turned into concentrated epidemic in February 2001 [3] (Figs 2, 3). In 2001 the prevalence of HIV among male IDUs was 12.9%, female IDUs – 13.7%, male detainees – 16.2%, female detainees – 17.3% and among pregnant women – 0.71%. Age group 20–24 was at highest risk of HIV infection (43.0% for men and 52.0% for women). HIV incidence rate was 107.8 for the country, 132.0 for the capital Tallinn and 572.1 for the most affected town Narva next to the border of Russian Federation. The rate of new HIV infections in Narva was highest in female 15–19 (50.0%) and male 20–24 (42.7%) age groups [2].

Start of the epidemic in Narva was favoured by a large number of IDUs in the area, connected probably with a difficult economic situation and high unemployment rate. The epidemic has since spread to other towns in the North-Eastern part of Estonia and later to capital city Tallinn. The rapid increase in intravenous drug use started in 1994–1995 with no decrease until now. IDUs are mostly heroin users, about 85% of them Russian-speaking and 75% male, aged 15 to 25. The estimated number of IDUs in the whole country is 12 000 – 17 000 [1]. Economic distress and reforms as well as political restructuring have had a radical impact on everyday life, as well as a negative impact on the health care system. This makes some communities especially vulnerable to outbreak of HIV infection [9].

Like with HIV infection, hepatitis B and C are most prevalent among drug addicts, especially among inmates (82.3% and 93.7% in 2000). We revealed also the high level of HBV and HCV prevalence among such high risk groups as health care workers (22.3% and 3.9%), hemodialysis patients (21.9% and 7.9%), and prison staff (20.0% and 12.5%), respectively [7] (Table 5). The same data for blood donors are in average 0.1% and 0.3%, respectively.

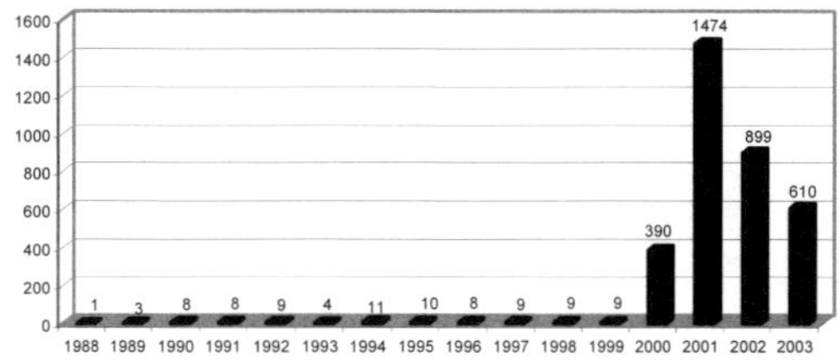

Figure 2. HIV infection cases in Estonia (1988–16.09.2003).

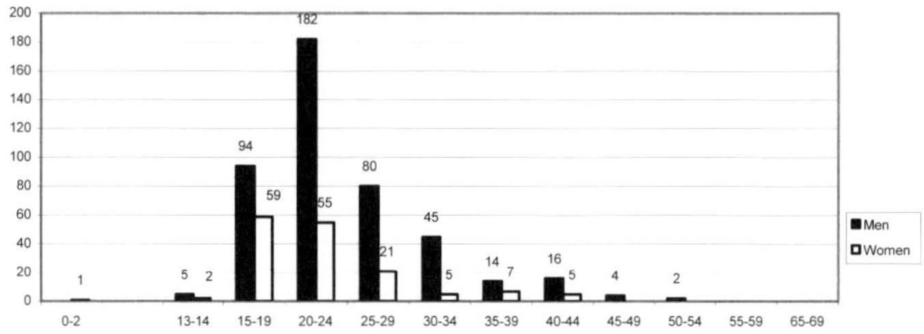

Figure 3. HIV infected by age and sexing year 2003.

Table 5. Infectivity with HBV and HCV in some risk groups of population in Estonia*.

Groups, years	HBV	HCV
Drug addict inmates, (1996–2000), n = 237	82.3% (12.7% HBsAg)	93.7%
Non-drug addict inmates (2000) n = 40	20.0% (10% HBsAg)	12.5%
Prison staff (2000) N = 34	20.6% (2.9% HBsAg)	2.9%
Health care workers (1995–1997), n = 510	22.3% (2.6% HBsAg)	3.9%
Hemodialysis patients (1994–2002), n = 256	21.9% (1.6% HBsAg)	7.9%

* Data of Virology Department, National Institute for Health Development.

Though hepatitis A is usually spread by food and water, the initial cases of the last outbreak of hepatitis A in 1998–1999 (Fig. 4) with 1084 cases in Northern and North-Eastern part of Estonia, was also connected with injecting drug use. A small group of young drug users in North-Eastern part of Estonia was assumed to be the source for the hepatitis A outbreak in early 1998 (data of Harjumaa-Tallinn and Ida-Virumaa Health Protection Services). The successive build-up of none-immune individuals after the previous outbreak in 1985 was likely reason for the rapid spread of hepatitis A into the general population.

Genetic analysis of hepatitis A virus strains revealed that despite of dominating subtype 1A before, the outbreak of HAV during 1998–1999 in Estonia was due to subtype IIIA, previously linked to addiction in Sweden during 1980's and in Norway at the end of 1990's [6].

The genetic analysis of Estonian hepatitis B virus strains revealed that they belonged only to two genotypes – D and A. Infection with genotype D was mainly associated with

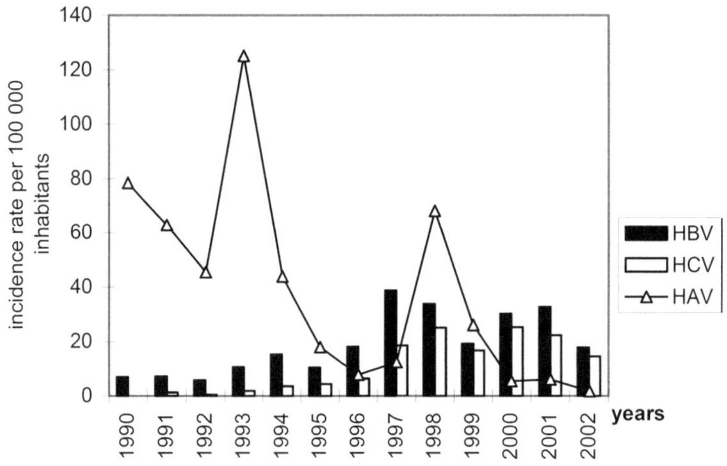

Figure 4. Incidence rate of viral hepatitis A, B and C per 100 000 inhabitants in Estonia (1990–2002).

IDU, sexual contacts and medical interventions [5]. HCV genotyping demonstrated that in Estonia the genotypes 1b and 3a seem to be more common among HCV infected persons [4,10].

Conclusion

High prevalence of drug use during last years resulted in the significant increase in morbidity with hepatitis B and C and concentrated HIV epidemic in Estonia.

Free of charge HBV vaccination in Estonia have covered till now only some population groups: health-care workers and medical students, adolescents (13 years old), and newborns. Slowing down the spread of viral hepatitis could be achieved both through immunization (HBV) of high risk persons and behavioral interventions to reduce harm and risk for HBV/HCV infections, and through appropriate medical management.

References

[1] Kalikov J, Kalikova N (2002) HIV and AIDS in Estonia. Ed. Minna Sinkkonen, *Stakes*, Finland.

[2] Kutsar K (2002) Emerging concentrated HIV epidemic in Estonia. XIV International AIDS Conference 2002, Barcelona, July 7–12. Abstract Book, vol. 2, WePeC6111:110.

[3] Priimägi L (2001) Beginning of HIV epidemic in Estonia. *Russian Journal of HIV/AIDS and related problems,* 5, 1:128.

[4] Tallo T, Lappalainen M, Tefanova V, Priimägi L (2000) Distribution of hepatitis C virus genotypes in patients with chronic hepatitis C in Northern Estonia. *Acta virologica,* 44 (3–4):175–178.

[5] Tallo T, Norder H, Tefanova V, Priimägi L, Magnius L (2002) Genetic characterization of HBV strains from Estonia. Book of Abstracts of the 5th Nordic-Baltic congress on Infectious diseases "Towards optimal diagnostics and management", St. Petersburg, 76–77.

[6] Tallo T, Norder H, Tefanova V, Ott K, Ustina V, Prükk T, Solomonova O, Schmidt J, Zilmer K, Priimägi L, Krispin T, Magnius L (2003) Sequental changes in hepatitis A virus genotype distribution in Estonia during 1994 to 2001. *J Med Virol,* 70 (2):187–193.

[7] Tefanova V, Tallo T, Priimägi L, Kikoš G, Krupskaja L (1999) Epidemiological aspects of viral hepatitis in Estonia. *Zn Mikrobiol Epidemiol Immunobiol,* 2:39–42.

[8] Tefanova V, Tallo T, Jaroslavtsev N, Priimägi L (2003) Viral hepatitis B and C among injecting drug users in prisons and visitors of anonymous consulting rooms during the dramatic increase of HIV infection in Estonia. J Clin Virol, 27 suppl. 1:16–72.
[9] Uusküla A, Kalikova N, Zilmer K, Tammai L, DeHovitz J (2002) The role of injection drug use in the emergence of HIV in Estonia. J Infect Dis, 6 (1):23–27.
[10] Zusinaite E, Krispin T, Raukas E, Kiiver K, Salupere R, Ott K, Ustina V, Zilmer K, Schmidt J, Sizemski L, Jaago K, Luman M, Ilmoja M, Prükk T, Ustav M (2003) Hepatitis C virus genotypes in Estonia. APMIS, 108 (11):739–746.

Emerging Biological Threat
G. Berencsi et al. (Eds.)
IOS Press, 2005

Epidemiology of Influenza A(H1N1) as One of Emerging-Reemerging Diseases

A.N. SLEPUSHKIN, E.I. BURTSEVA, L.N. VLASSOVA and V.T. IVANOVA
D.I. Ivanovsky Research Institute of Virology Russian Academy of Medical Science

Influenza, which are yearly circulating in the cities registered from the last decade of 19th century. The same time it is an emerging threat, because influenza A subtypes periodically are changing their genetic and antigenic structures by reassortment of the RNA segments called "shift" variations.

Especially peculiar epidemic behavior have been observed in the case of influenza A/H1N1 infections during the last 40 years. Variants circulating in the world during 1947–57 disappeared due to the emergence of subtype influenza A/H2N2. The latter has been also eliminated by the emerging next pandemic strain influenza A/H3N2 in 1968–69.

Influenza A/H1N1 reemerged, however, in 1977 causing pandemic among people mostly younger then 30 years of age. Elder cohorts had been protected probably by residual immunity since it happened that the antigenic structure of A/USSR/90/77-like A/H1N1 viruses possessed nearly identical antigenic properties than those circulating in 1950th.

There were four epidemics of influenza A/H1N1 in Russia after 1977–78 (in seasons 1981–82, 1984, 1986–87 and 1989). Influenza A/H1N1 viruses continued to undergo antigenic "drift" in comparison to the reference strains from A/USSR/90/77-like strains to A/Singapure/6/86-like variants. The spread of influenza A/H1N1 viruses did not prevent the circulation of A/H3N2-caused influenza A epidemics. The epidemics caused by influenza A/H1N1 variants alternated with epidemics caused by influenza A/H3N2.

The A/H1N1) viruses disappeared from open epidemic circulation for 5 years from summer 1989 till summer 1995. The A/H1N1 viruses caused mostly local outbreaks among schoolchildren and youth during the epidemic season 1995–96. The isolated influenza A/H1N1/1995 strains were A/Texas/36/91-like. Their hemagglutinins (HA1) were different in 8–10 amino acid positions from those of strains isolated between 1986–1989. The antigenic "drift" during 1990–1995 moved into a different direction than those isolated from 1950 to 1956.

After 1996 no open epidemics were caused by influenza A/H1N1 again for 5 years. The influenza A/H1N1 reemerged in the epidemic season in 2000–2001. Antigenic modifications of these variations are discussed.

1. Introduction

Epidemiological data of influenza viruses, which are circulating in the cities are known from the last decade of 19th century.

The influenza A/H1N1 is a virus of peculiar epidemic behavior. Its variety was well known as Influenza A prime. At first it circulated among people during 1948–57 and later a new pandemic influenza A/H2N2 eliminated its epidemic circulation [1]. Then the influenza A/H1N1 subtype reemerged in 1977, when there was a pandemic burst mostly among people up to age 30 [2,3].

This was probably connected with the antigenic structure of type A/H1N1 strains appeared in 1977 which have been nearly identical to those strains circulated in the 50th. Most people born before 1950 possessed residual immunity to them [3,4]. There were four epidemics of influenza A/H1N1 in Russia and other countries after 1977. These occurred in the seasons 1981–82, 1984, 1986–87, 1989. During that time influenza A(H1N1) viruses continued to undergo antigenic variation from A/USSR/90/77-like reference strains to A/Taiwan/01/86 or A/Singapore/6/86-like. Later on they disappeared again from epidemic circulation. During the next 5 years up to the summer of 1995 these viruses were isolated in Russia and other countries only from sporadic cases of influenza like illnesses (ILI).

In the summer of 1995 the influenza A/H1N1 subtype re-emerged at an epidemic level in Australia first with an antigenic structure similar to A/Singapore/6/86 or A/Texas/36/91 [5,6].

2. Materials and Methods

The Russian Department of Sanitary and Epidemiological Surveillance of the Ministry of Health conducts morbidity monitoring. Reports are prepared for the Federal Center of Influenza (Saint-Petersburg), to the Center of Influenza Ecology and Epidemiology at the Ivanovsky Institute of Virology, Moscow and Regional Centers for Epidemiological Surveillance. The expected number of Influenza and ARD are plotted for each region and epidemic thresholds are calculated. When the numbers of Influenza and ARD morbidity exceed this threshold, an Influenza epidemic is deemed to be under way.

The isolation and propagation of influenza viruses are carried out on chicken embryos or MDCK tissue culture. The antigenic structure was studied in neutralisation tests with immune sera of white rats, prepared at the Influenza Etiology and Epidemiology laboratory. Hemagglutination-Inhibition (HI) test were performed as recommended by World Health Organization [7].

The standard methods of HI test were used for titration of patients paired sera that had been previously treated with receptor-destroying enzyme [7]. The fourfold rise in titers of antibodies were considered as diagnostically positive.

3. Results

3.1. Epidemiologic Character of the Two Re-Emergence if Influenza A/H1N1

The dynamic of influenza epidemic in 1995–96 in Russia is presented on Fig. 1. Dates of the beginning of the epidemic have been declared, when the number of ARD/ILI exceeded the epidemic thresholds in the administrative regions. The dates of the ends of the local epidemics have been defined when the ILI/ARD activity decreased below the epidemic threshold. The number of administrative territories affected by the epidemic have been registered weekly. According to the data of the Influenza Surveillance Center the this epidemic has emerged in two separate regions of Russia simultaneously. First of all it appeared in Western Kaliningrad region on week 46. But at the same time the onset of the epidemic was registered also in the center of the European Russia (e.g. Tver, Moscow City and Moscow Region, Tula, Bryansk and Tambov). Another probably a third independent focus of influenza infections has developed at the Volga River and in the South-Western Ural in a big industrial city Ufa one week (48 week) later.

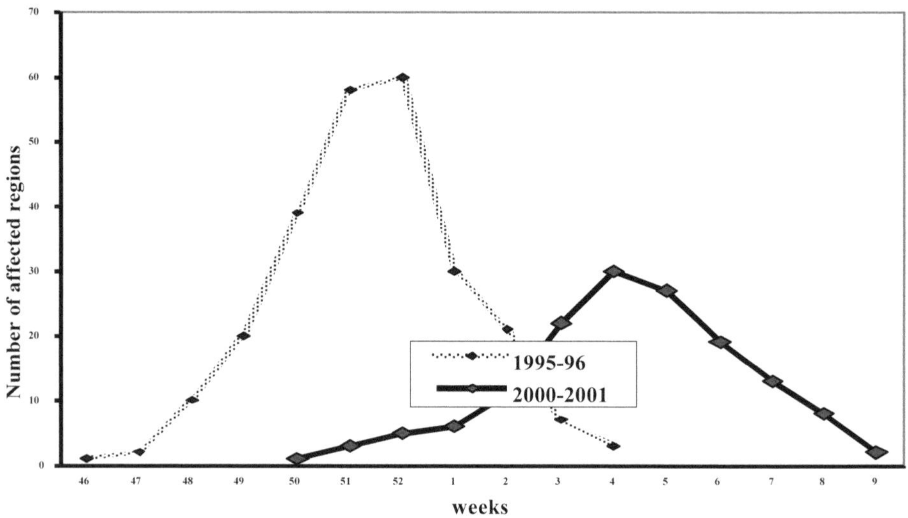

Figure 1. Influenza A epidemics in 1995/96 and 2000/2001 in Russia.

Then the influenza epidemic spread from Central Russia to the North-East, Southern and North-Western European Russia. After that it spread into the Eastern and Western Siberia and other Far East regions during 50th, 51st and 52nd weeks of 1995.

In Moscow influenza and ARD morbidity came down below the epidemic threshold from the first week of 1996. The morbidity of schoolchildren, however, increased again from the third week of January when they returned back to schools after the New Year and Christmas holidays. The same trend was registered in four other administrative territories, too. That morbidity values, however, still remained below the epidemic levels during the second week of 1996.

The length of epidemics in different administrative regions was 3 to 5 weeks.

The rate of morbidity was 2837 per 100 000 inhabitants in Moscow at the peak of epidemic (week 49). This level can be practically compared with the morbidity level registered in 1989 namely 2773 cases per 100 000 people.

Over 50% of the patients infected during the first weeks of epidemic were children under 14 years of age. The number of adults involved in the epidemic was considerably higher at its peak and came up to 62%. At the end of the epidemic the number of infected children increased again up to 55%.

As far as the etiology of the epidemic is concerned all three influenza viruses were involved but mainly influenza viruses A/H1N1 and A/H3N2 were circulating.

Influenza A/H1N1 infections were diagnosed in the Moscow Region and Moscow, in cities Yaroslavl, Bryansk, Vladimir, Penza, Kursk, Lipetsk, Tambov, Novgorod, and Ekaterinburg. Influenza A/H3N2 – was detected in all above mentioned regions as well as in Kaliningrad, Tver, Tula, Saratov, Kirov, Vologda, Karelia, Orenburg, Novosibirsk, Khabarovsk. The influenza B virus was diagnosed in Moscow, Bryansk, Vologda, Novgorod, Karelia, Kirov, Lipetsk, Orenburg, and Novosibirsk.

In order to control the involvement of influenza A/H3N2 and A/H1N1) in the epidemic several sera of patients suffering previously from ILI in Moscow, Moscow Region, Novgorod, Vladimir and Yaroslavl have been tested. The etiology of local outbreaks in schools and conscripts units were compared with the viral etiology measured in the general population. In Moscow the etiological agent was influenza A/H1N1 in the case of two outbreaks

Table 1. Results of hemagglutination-inhibition tests with sera of influenza and ARD patients of 1995–1996.

City	Type of population	Number of patients	Number of seroconversions to the viruses		
			A(H1N1)	A(H3N2)	B
Moscow	Five local outbreaks in schools and conscripts units	76	42	2	
	General population	29	4	11	1
Vladimir	General population	30	7	16	3
Novgorod	General population	27	4	10	2
Yaroslavl	General population	26	9	12	3

Table 2. Antigenic properties of A(H1N1) strains, isolated in Russia during epidemic 1995–1996 and 2000–2001.

Epidemic Season	Virus isolation			Antigenic structure like:
	Regions	Number of strains		
		In region	Total	
1995–1996	Moscow Lipetsk Vladimir	5 2 1	8	A/Texas/36/91
2000–2001	Moscow Lipetsk Vladimir Novgorod Yaroslavl Nizhny Novgorod	10 13 3 10 1 5	42	A/New Caledonia/20/99

in schools and three in conscripts units (Table 1). At the same time among people of general population from 29 patients 10 were affected by influenza A/H3N2 and only 4 by influenza A/H1N1. Children affected by a local influenza outbreak in town Egoryevsk in the Moscow Region suffered also from influenza A/H1N1 which was confirmed by isolation of five strains. At the same time citizens in Vladimir, Yaroslavl and Novgorod were suffering from influenza/H3N2 since most of them had 4-fold rise of antibody titre to A/H3N2/Iohannesburg/33/94 virus but several of them suffered from influenza A/H1N1.

From throat swabs of patients influenza A/H1N1 virus strains were isolated on chicken embryos and MDCK cells in the Moscow Region, Vladimir, Lipetsk and Ekaterinburg. Influenza A/H3N2 was isolated in Khabarovsk and Minsk (Belarus) but, influenza B virus – in St. Petersburg.

3.2. Antigenic Properties of the Isolated Influenza A/H1N1 Viruses from 1995/1996

The Influenza Ecology and Epidemiology Center working at the Ivanovsky Research Institute of Virology studied antigenic structures of eight A/H1N1 strains isolated from patients from the Moscow Region, Vladimir and Lipetsk (Table 2). All of them proved to be A/Taiwan/1/86- and A/Texas/36/91-like. Two influenza B strains isolated in Saint Petersburg were influenza B/Beijing/184/94-like.

Table 3. Isolation of Influenza viruses in Russia during 2000–2001 epidemic season.

Type and Subtype of Viruses	City and region	Number of isolation	Antigenic structure like WHO etalon	Method of isolation
A(H1N1)	Moscow and Moscow region	10	A/New Caledonia /20/99	Chicken embryos (CE) – 2 strains MDCK – 8 strains
Type B	Moscow Moscow	3 6	B/Sichuan/379/99 B/Yamanashi/166/98	MDCK
A(H1N1)	Lipetsk and Lipetsk region	13	A/New Caledonia /20/99	MDCK
A(H1N1)	Novgorod and Novgorod region	10	A/New Caledonia /20/99	CE
A(H1N1)	Vladimir and Vladimir region	3	A/New Caledonia /20/99	CE
A(H1N1)	Yaroslavl	1	A/New Caledonia /20/99	MDCK
A(H1N1)	Khabarovsk	1	A/New Caledonia /20/99	CE

3.3. Characteristics of the Influenza A/H1N1 Epidemic from 2000/2001

Influenza A(H1N1) did not evoke epidemics in the world during next 5 years. The next epidemic occurred in the season 2000–2001. According to the data of the Ministry of Public Health and Influenza Surveillance Centers the epidemic increase of morbidity of ILI began in the Yaroslavl Region in the 50th week of 2000 (Fig. 1). There were three districts with local epidemic foci registered among schoolchildren. The epidemic thresholds of influenza morbidity among schoolchildren were overtaken in ten cities of European part of Russia during week 51 of 2000. Influenza viruses A/H1N1 were isolated from nasal washings of two patients in Moscow (1-st Hospital of Infectious Diseases) by the Ecology and Epidemiology Center of our Institute at the end of December.

The New Year, Christmas Festivals and schoolchildren vacation influenced ARD morbidity as before. Its level decreased during weeks 52nd and 1st in 2001. The epidemic levels of influenza morbidity were registered in 8 cities of the European part of Russia and West Siberia on the 2nd week of January, 2001. That was the real beginning of the epidemic.

The epidemic thresholds of influenza morbidity among the general population were overtaken in 13 cities. Among the schoolchildren, however, the epidemic thresholds were overtaken in 22 cities during the 3rd week of 2001. Twenty-six influenza A/H1N1 strains were isolated during that period.

Lastly the epidemic thresholds of influenza morbidity were overtaken among people of Khabarovsk city during week six of 2001. The gradual decrease of morbidity began from week seven in 6 cities of European part of Russia. But it was still high in 5 Siberian cities. The epidemic was over in 12 cities in week nine. Further decreases of morbidity took place during weeks 10 and 11 of 2001.

The Center of Influenza Ecology and Epidemiology in our Institute isolated or studied antigenic structure of 47 epidemic strains (Table 3). The number of isolated A/H1N1 strains was 38. All influenza A/H1N1 strains were A/H1N1/New Caledonia/20/99-like. Also 9 influenza B strains were isolated during the season. From influenza A/H1N1 strains 21 were isolated on MDCK cells and 17 on chicken embryos. All influenza B strains were isolated on MDCK cells. Investigators of the WHO Influenza Center in CDC, Atlanta, USA sequenced some isolated strains from 2001. The analyses indicated 12–13 changes in amino acid positions in comparison to the reference strain A/H1N1/Beijing/262/95 and only 3–4

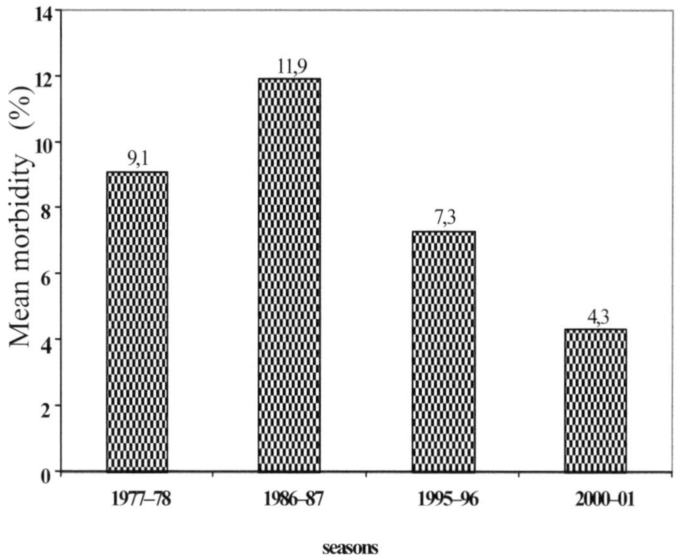

Figure 2. Mean morbidities of influenza A/H1N1 epidemics in Russia.

changes in comparison to the last reference strain A/H1N1/New Caledonia/20/99.Five regional laboratories isolated also influenza A/H1N1 strains. About 50 others were isolated by the Federal Influenza Center, St. Petersburg and 11 by its regional laboratories.

The virological data and serological studies confirmed that the main etiological factor of this epidemic was influenza A/H1N1. Influenza B had only secondary role especially at the end of epidemic season in several regions (Nizny Novgorod, Khabarovsk, Udmurt Republic). There were information about some cases of Influenza A/H3N2, but this virus was not isolated during the influenza season in Russia.

The length of the epidemic in various regions was different. From 3 weeks (Omsk Region) to 6,5 weeks (Ivanovsk Region). The level of morbidity was also variable from 4,1% of population (Omsk Region) up to 7,6% (Ivanovsk Region). The morbidity among children was 3–4 times higher: from 14,7 to 24,9% of the corresponding cohorts.

4. Discussion

As it turned out from results of epidemiology surveillance the 1995–96 epidemic in Russia began from Western (Kaliningrad) and Central (Tver, Moscow, Tula) regions. Etiological factors were Influenza A(H1N1) and A(H3N2) viruses. The Influenza A(H1N1) unlike in previous five years caused during this epidemic not only sporadic cases but local outbreaks among schoolchildren and conscripts. Two factors may be taken in consideration for explanation of this change in epidemic behavior of Influenza A(H1N1).

One of them is decrease of the immunity level to Influenza A(H1N1) among population during previous five years especial among children and young people who had no or only short contact with this virus previously. The second factor might be change in biological properties of virus for example to become more temperature stable.

The numbers of mean ARD morbidity are shown in Fig. 2 of the epidemics in 1977–78, 1984, 1995–96 and 2000–2001. The level of morbidity decreased significantly from 11,9%, 7,3% to 4,3%.

To conclude we would like to point out the following: activity of this virus is decreasing. That was confirmed by the fact that epidemics of influenza A/H1N1 for the last 10 years reemerged only once per every 5 years and their intensity gradually were going down.

References

[1] Slepushkin A.N. The effect of previous attack of A1 influenza on susceptibility to A2 virus during the 1957 outbreak // Bull. WHO. – 1959. – 20. – P. 297–301.
[2] Zhdanov V.M., Lvov D.K., Zakstelskay L.Ya et al. Return of epidemic A(H1N1) influenza virus // Lancet. – 1978. – N 2. – P. 294–295.
[3] Slepushkin A.N., Obrosova-Serova N.P., Schastny E.I. et al. Immunostructure of Moscow population to strains of Influenza viruses A(H3N2), A(H1N1) and B, circulated in 1977 (in Russian) Voprosy virusologii (Questions of virology). – 1979. – N 2. – P. 175–180.
[4] Kendal A.P. Epidemic Implications of Changes in the Influenza virus genome // Am. Journal of Med. – 1987. – Vol. 82. – P. 4–14.
[5] WHO Wkly Epidem. Rec. – 1995. – N 17. – P. 122.
[6] WHO Wkly Epidem. Rec. – 1995. – N 23. – P. 167.
[7] Palmer D.F., Dowdle W.R., Coleman M.T., Schild G.C. Advanced laboratory techniques for influenza diagnosis // U.S. Department of Health, Education and Welfare, Center for Disease control, Atlanta, GA 30333/ – 1975. – N 1. – P. 26–62.

Emerging Biological Threat
G. Berencsi et al. (Eds.)
IOS Press, 2005

West Nile and Other Emerging-Reemerging Viruses in Russia

D.K. LVOV[a,b], A.M. BUTENKO[a], V.L. GROMASHEVSKY[a], M.Yu. SHCHELKANOV[a], A.I. KOVTUNOV[a], K.B. YASHKULOV[d], A.G. PRILIPOV[a], R. KINNEY[e], V.A. ARISTOVA[a], A.F. DZHARKENOV[c], E.I. SAMOKHVALOV[a], H.M. SAVAGE[e], I.V. GALKINA[a], P.G. DERYABIN[a,b], B.Ts. BUSHKIEVA[d], D.J. GUBLER[e], L.N. KULIKOVA[c], S.K. ALKHOVSKY[a], T.M. MOSKVINA[a], L.V. ZLOBINA[c], G.K. SADYKOVA[a], A.G. SHATALOV[a], D.N. LVOV[a,b], V.E. USACHEV[a] and A.G. VORONINA[a]

[a]*D.I. Ivanovsky Institute of Virology RAMS, Moscow, Russia*
E-mail: dk_lvov@mail.ru, Tel.: +7 (095) 190-2872, Fax: +7 (095) 190-2867
[b]*I.M. Sechenov Moscow Medical Academy, Moscow, Russia*
[c]*State Center of Sanitary-Epidemiological Inspection, Astrakhan, Russia*
[d]*State Center of Sanitary-Epidemiological Inspection, Elista, Russia*
[e]*Division of Vector-Borne Infection Diseases, National Center for Infectious Diseases, Centers for Disease Control and Prevention, Fort Collins, Colorado, USA*

Studies of the interactions of vertebrates, viruses and arthropod vectors of these viruses were monitored in terms of different ecological groups of viruses transmitted by mosquitoes and ticks in Northern Eurasia in an area encompassing more than 15 million km^2. About 90 viruses were isolated, including 24 new to science. Newly recognized infections of vertebrates, including humans, were described. Many unusual epidemic situations were analysed. Permanent efforts were established to prevent bioterrorist activities and their consequences. Extensive epidemic outbreaks of West Nile fever (WNF; i.e., fever caused by West Nile virus) and Crimean-Congo hemorrhagic fever (CCHF) with unusual high mortality appeared in the last four years in southern Russia. Infection rates in humans, domestic and wild animals, mosquitoes and ticks from natural and anthropogenic biocenosis had been determined. CCHF virus strains were phylogenetically similar to strains isolated in this area 35 years ago but different from Central-South-Asian and African strains. Before the outset of the current emergence of epidemic WNF, three genetic variants of this virus had been isolated in USSR, two African and one Indian. Phylogenetic analysis of complete genome sequences of epidemic strains demonstrated considerable similarity to strains from USA and Israel and differences from strains isolated in the same USSR areas 20–30 years before. In addition to strains of genotype 1, we isolated strains of second and third lineages and a strain of a fourth genetic variant. Nucleotide differences of these strains from all three genotypes was about 30%. The emerging WNF situation in Russia for the last 4 years probably has been the result of not only by natural and social factors but also by evolution of the virus.

Introduction

In the former Soviet Union there was a functioning and vigorous anti-terrorist system, similar in purpose to CDC's Epidemiological Intelligence Service [9,15]. The purpose and methodological approach provided ecological evaluations of various regions, with collection of field materials in longitudinal sections about 2,000 km long throughout the entire territory of northern Eurasia. In this way a land mass of more than 15 million km^2 was studied. Twelve zones were distinguished, each passing through unique ecosystems, including arctic, tundra, taiga, deciduous forest, steppe, and deserts within 18 physico-geographical areas. About 90 different viruses were isolated, of which 24 were new to science. The etiological role of these viruses in human disease was evaluated, with four of them causing previously unrecognized illnesses. As well, the potential for emergence of epidemic situations in various landscape zones was determined. Predicting areas of emergence of new viruses also was done [18,20–22] and all this was part of a permanent system to prevent bioterrorism events and consequences for the foreseeable future [17].

Within the past four years there have been examples of re-emerging epidemic situations in the south of Russia, including extensive outbreaks of Crimean-Congo hemorrhagic fever (CCHF), West Nile fever (WNF; i.e., fever caused by West Nile virus; WNV) and perhaps Dhori virus (DHOV) [17,23–25]. WNF endemic territories in southern Russia have been recognized since the first isolation of this virus was made in the north of the Caspian Sea basin region 40 years ago. Sporadic morbidity and small outbreaks were observed in southern regions of the Soviet Union practically every year. Serological results indicated, however, that the WNV is present even in the Karpathian Basin [11]. Antibody prevalence within the population of the USSR was known, and highest indices were found in southern Russia [18–22]. Therefore the epidemic of 1999, originally diagnosed by local specialists as due to enterovirus infections, was not unexpected, although the epidemic had unique features, principally the magnitude of the epidemic. Laboratory-verified cases alone were more than 500 in 1999, and the number of infected patients exceeded 200,000, according to our calculations on the basis of the results of serological examination before and after outbreak [16,17,24,33]. Mortality was also unusually high, about 10% [17,23,24]. Interestingly, a simultaneous and similar epidemic outbreak occurred in the New York City area [6,13,27], which has spread to essentially all of the continental United States and which continues [10,12,26]. The origin of that virus was Northern-Africa. To the south of Europe, first of all through the deltas of large rivers, such as the Rhone, Danube, and Volga, constant introduction of viruses has taken place via migration routes of birds, with subsequent introduction of viruses into local populations of birds and mosquitoes [11,19,20]. The main natural focus in our country is at the Volga River delta. Natural and social factors, as is case with CCHF, were of importance in the development of the epidemic, but we also had to consider the possibility that what had appeared was a viral population with unusual genetic characteristics. By the outset of the epidemic three genetic variants of WNV were known: two African and one Indian [3–5,13,14,30,32,34]. Introduction of the agent into the Volga River delta takes place basically from Africa, where more than 90% of birds nesting or flying over the delta are members of overwintering species, with some proportion of birds having overwintered in the Indian subcontinent [20]. We considered it useful to collect field materials in this region and to analyse them by virological, serological and genetic methods so that we might answer the question: What species of birds, mosquitoes, and domestic animals are involved in WNV circulation and what are their genetic properties? To answer these questions was the rationale for our studies of the genomes of epidemic strains of WNV isolated from different sources in various years in the south of Russia and other regions of the USSR; we then intended to compare our data with data regarding WNV strains from Romania, Israel, India, Africa, and USA.

Materials and Methods

The delta of the Volga River consists of three basic areas with unique ecosystems features. The Low-delta is located at the edges of the Caspian Sea and is characterized by extensive, exposed water spaces. The depths of the sea in these areas usually do not exceed 1.0–1.5 meters, which promotes mass mosquito propagation and provides simultaneous nesting opportunity for waterfowl and shore birds. The Middle-delta is more distant from the sea and consists of powerful current and shallow lone ecosystem with reeds and shrubs. Here natural ecosystems are combined with anthropogenic areas around a number of settlements, where inhabitants keep cattle, sheep, camels, and horses. The High-delta adjoins the Volga-Akhtuba lowland as well as semi-desert and desert. Cities including Astrakhan are located here. Some species of birds that are usual to Middle-delta also occur in this belt, closely contacting with domestic animals and birds of human habitats. Most morbidity caused by WNV occurs in the High-delta belt. The virus could be introduced from the delta to the Volga-Akhtuba area and probably further north to steppes of low Volga belt, with large cities and mostly anthropogenic biocenosis. Field materials (brain and spleen from birds and mammals; mosquitoes; ticks; blood from patients; sera from domestic animals) were collected in August of 2001–2002 (Tables 1, 2) and immediately put into liquid nitrogen. Mosquitoes were collected by hand-held, battery- powered aspirators and in light traps. For virus isolations we tested spleen and brain from birds and mammals and pools of mosquitoes, 100 per pool, which were tested for virus using VeroE6 and pig kidney cells. In some cases we also used 3–4 day-old suckling mice, which were inoculated intracranially with 0.01ml of clarified tissue suspension; when necessary, one blind passage was made 6 to 7 days after inoculation. All manipulation with mice satisfied international requirements for occupational health and safety in the care and use of research animals. When mice began to show signs of illness, brain tissue was removed and used for identification by RT-PCR, enzyme-linked immunosorbent assay (ELISA), and hemagglutination inhibition test (HIT) with immune ascite fluids (IAF) of mice to viruses previously isolated in this region: Tahyna and Batai (*Bunyaviridae*, *Orthobunyavirus*), CCHF (*Bunyaviridae, Nairovirus*), Bhanja (*Bunyaviridae*), Sindbis (*Togaviridae*, *Alphavirus*), WNV (*Flaviviridae*, *Flavivirus*), and Dhori (*Orthomyxoviridae*, *Thogotovirus*) [34]. Neutralization testing (NT) was done by the micro method using pig kidney cells, with a single dilution of sera or IAF and 10-fold dilutions of virus.

WNV RNA detection was extracted from 0.2–0.3 ml sample using a guanidine-isothiocyanate-phenol-chloroform method. cDNA synthesis was performed using WNV-specific antisense 23-mer primer GTGCACCAACAGTCGATGTCTTC, and AMV-derived reverse transcriptase (Promega). Amplification of WNV RNA-specific sequence was carried out by PCR using primers derived from a well-conserved 5'-region of the WNV genome. Briefly, the first-round PCR amplified a sequence with primers 1F/518R (25 cycles of 94°C – 20 sec/55°C – 20 sec/72°C – 30 sec), and the second-round PCR amplified an inner sequence of 487 bp in length with primers 26F/518R (35 cycles of 94°C – 20 sec/60°C – 20 sec/72°C – 30 sec). Primers: 23-mer 1F – AGTAGTTCGCCTGTGTG AGCTGA; 23-mer 518R – TCGGTAGCATTTACCGTCATCAT; 22-mer 26F – AACTT-AGTAGTGTTTGTGAGGA. Both strands of the PCR product were then sequenced on a 377 automated sequencer (Applied Biosystems, USA). Sequences have been composed with Lasergene SeqMan computer program (DNAstar, USA). Consensus sequences then were aligned by using the program MegAlign (DNAstar, USA). WNV strains sequences tested in this study are presented in Table 3.

Table 1. Virus isolation* and WNV RNA detection, field materials collected 2001–2002.

Object	Order	Family	Genera	Species	RT-PCR results: Volga-delta: Number of investigated	Volga-delta: Positive No	Volga-delta: Positive %	Volga-Akhtuba: Number of investigated	Volga-Akhtuba: Positive No	Volga-Akhtuba: Positive %	Total: Number of investigated	Total: Positive No	Total: Positive %
Birds	Podicipidiformes	Podicipitidae	Podiceps	2 species	7	0	0.0	12	0	0.0	19	0	0.0
	Anseriformes	Anatidae	Anas	5 species	22	0	0.0	19	0	0.0	41	0	0.0
	Gruiformes	Rallidae	Fulica	F. atra	158	18	11.4	3	0	0.0	161	18	11.2
	Pelecaniformes	Phalacrocoracidae	Phalacrocorax	2 species	112[*1]	23	20.5	3	0	0.0	115	23	20.0
	Ciconiiformes	Ardeidae	4 genera	6 species	72	5	6.9	10	0	0.0	82	5	6.1
		Threskiornithidae	Plegadis	P. falcinellus	31	0	0.0	0	0	0.0	31	0	0
	Charadriiformes	Charadriidae	3 genera	3 species	43	1	2.3	1	0	0.0	44	1	2.3
		Laridae	2 genera	6 species	183	9	4.9	39	1	2.6	222	10	4.5
	Passeriformes	Corvidae	4 genera	5 species	139[*2]	38	27.3	53	2	3.8	192	40	20.8
		4 families	4 genera	4 species	89	4	4.5	13	0	0.0	102	4	3.9
	Falconiformes	2 families	2 genera	2 species	11	0	0.0	2	0	0.0	13	0	0.0
	Cuculiformes	Cuculidae	Cuculus	C. canoris	5	1	20.0	0	0	0.0	5	1	20.0
	Caraciiformes	2 families	2 species	2 species	11	1	9.1	1	1	(100)	12	2	16.6
	Galliformes	Phasianidae	Perdix	P. perdix	7	1	14.3	1	0	0.0	8	1	12.5
	Columbiformes	Columbidae	Columba	C. livia	11[*3]	2	18.2	5	0	0.0	16	2	12.5
Total	12 order	20 families	29 genera	42 species	901	103	11.4	162	4	2.5	1063	107	10.1
Mammals	Lagomorpha	Leporidae	Lepus	L. europaeus	8[*4]	2	25.0	5	0	0.0	13	2	15.4
	Insectivora	Erinaceidae	Hemiechinus	H. auritus	3	0	0.0	2	0	0.0	5	0	0.0
	Rodentia	Dipodidae	Allactaga	A. elater	1	0	0.0	6	0	0.0	7	0	0.0
			Alactagulus	A. acontion	1	1	(100)	0	0	0.0	1	1	(100)
Total	3 order	3 families	4 genera		13	3	23.1	13	0	0.0	26	3	11.5
Ticks	Parasitiformes	Ixodidae	Hyalomma	H. marginatum	160[*5]	nt	Nt	0	0	0.0	160	nt	nt
Mosquitoes	Diptera	Culicidae	Culex	C. pipiens	(s) 11400	2	0.018	300	1	0.333	11700	3	0.026
				C. modestus	(s) 6700	2	0.030	2900	1	0.034	9600	3	0.031
					(n) 13000	0	0.0	0	0	0.0	13000	0	0
			Anopheles	A. messeae	(s) 15800[*6,7]	10	0.063	2600	2	0.077	18400	12	0.065
				A. hyrcanus	(s) 6900	5	0.072	0	0	0.0	6900	5	0.072
					(n) 17000	2	0.012	0	0	0.0	17000	2	0.012
			Coquillettidia	Coq. richiardii	(n) 12300	6	0.049	0	0	0.0	12300	6	0.049
Total	1 order	1 family	3 genera	5 species	(s) 40800	19	0.047	5800	4	0.069	46600	23	0.049
					(n) 42300	8	0.019	0	0	0.0	42300	8	0.019
					∑ 83100	27	0.033	5800	4	0.069	88900	31	0.035

* isolated strains:

[1] 3 WNV strains from Phalacrocorax carbo; [2] 4 WNV strains: 1 from Corvus frugilegus, 2 from Corvus corone, 1 from Pica pica; [3] 1 WNV strain from Columba livia; [4] 1 DHOV strain – from Lepus europaeus;

[5] 7 strains from H. marginatum: 2 – WNV, 3 – DhoV, 1 – CCHFV, 1 – DhoV and CCHFV mix; [6] 19 Batai virus strains from Anopheles messeae (infection rate 0.039 %). [7] 2 WNV strains from Anopheles messeae. (s) – anthropogenic biocenosis; (n) – natural biocenosis. Additionally one strain of CCHFV isolated from blood of patient in Volga-Akhtuba ecosystem.

Table 2. Serological examination of domestic animals in Volga-delta and Volga-Akhtuba, 2001–2002.

Animal	Ecosystem	NT			HIT		
		Number investigated	Positive results		Number investigated	Positive results	
			number	%		number	%
Cattle	Arid landscapes	488	18	3,7	488	11	2,3
	Volga-delta	88	28	31.8	373	31	8.3
	Volga-Akhtuba	54	5	9.3	150	15	10.0
	Total:	630	51	8.1	1011	57	5.6
Sheep	Arid landscapes	160	12	7.5	160	7	4.4
	Volga-delta	91	15	16.5	332	10	3.0
	Volga-Akhtuba	28	9	32.1	188	2	1.1
	Total:	279	36	12.9	680	19	2.38
Camels	Arid landscapes	17	0	0	17	0	0
	Volga-delta	24	2	8.3	76	9	11.8
	Volga-Akhtuba	38	6	15.8	83	2	2.4
	Total:	79	8	10.1	176	11	6.3
Horses	Arid landscapes	10	1	10	10	0	0
	Volga-delta	nt*	nt	nt	68	15	22.1
	Volga-Akhtuba	nt	nt	nt	17	0	0.0
	Total:	10	1	10	95	15	15.8
	Arid landscapes	675	31	4.6	675	18	2.7
	Volga-delta	203	45	22.2	849	65	7.7
	Volga-Akhtuba	120	20	16.7	438	19	4.3
Total		323	65	20.1	1287	84	6.5

* "nt" indicates "not tested".

Results

Among 38 strains isolated from field materials collected in Volga-delta and Volga-Akhtuba ecosystems in 2001–2002, 12 were of WNV, 3 of CCHFV, 4 of DHOV, and 19 of Batai virus (Table 1). All strains of WNV were isolated in the Volga delta. All strains in natural habitats were obtained from cormorants and in anthropogenic biocenosis – from *Anopheles messeae* mosquitoes, *Corvidae* birds (mostly crows) and, collected from them, pre-imago *Hyalomma marginatum* ticks. CCHFV strains were isolated from patient in Volga-Akhtuba ecosystem as well as from larva and nymphs of *H. marginatum* collected from a hare in anthropogenic biocenosis of the Volga delta ecosystem. From the same hare and ticks DHOV strains were also isolated. One was a mixture of CCHFV and DHOV. Batai virus was isolated from *An. messeae* mosquitoes in anthropogenic biocenosis (Table 2). WNV was detected by RT-PCR of tissues from birds, mostly those collected in the Volga delta and most often among cormorants, coots, herons, gulls, and terns in natural biocenosis. In anthropogenic biocenosis positive results were obtained among ground-feeding birds, especially among corvids, and from mammals (Table 2).

Table 3. West Nile virus strains used for phylogenetic study*.

#	Strain	Year of Isolation	Source of isolation	Place of isolation
1	Ast63-94-ticks	1963	Ticks *Hyalomma marginatum*	Russia, Volga delta
2	LEIV-Az67-1640-nathatch	1967	Bird *Sitta europaea*	Azerbaijan, Lencoran lowland
3	LEIV-Az67-1628-thrush	1967	Bird *Turdus merula*	Azerbaijan, Lencoran lowland
4	LEIV-Az70-72-ticks	1970	Ticks *Ornithodoros capensis*	Azerbaijan, Island in Caspian Sea
5	LEIV-Tur73-2914-ticks	1973	Ticks *Hyalomma detritum*	Turkmenistan, Uzboi valley
6	LEIV-Taj77-1304-tern	1977	Bird *Sterna albifrons*	Tajikistan, Pyanzh valley
7	LEIV-Ukr80-3266-rook	1980	Bird *Corvus frugilegus*	pre-Carpathian plain, Ukraine
8	LEIV-Krnd88-190-ticks	1988	Ticks *Dermacento marginatus*	N-W Caucasus, mountain valley
9	LEIV-Vlg99-27889-human	1999	Human brain (section material)	Russia, Volgograd, low Volga
10	LEIV-Vlg00-27924-human	1999	Human blood	Russia, Volgograd, low Volga
11	Ast99-901-human	1999	Human blood	Russia, Astrakhan, Volga delta

* WNV strain sequences from GenBank used for phylogenetic study: Eg101 (Egypt, 1951), Ig2266 (India, 1955), NC-001563 (Nigeria), Italy 1998 (Italy, 1998), RO97-50 (Romania, 1997), IS-98 STD (Israel, 1998), Connecticut (USA, 1999), NY-99 eqhs (USA, 1999), HNY-1999 (USA, 1999), NY-99 flamingo (USA, 1999).

Investigation by RT-PCR of mosquitoes showed they were involved in virus circulation among all predominant species in anthropogenic biocenosis, including *An. hyrcanus, Culex pipiens, Cx. modestus*, and *An. messeae*, and in natural biocenoses *Coquilettidia richiardii* and *An. hyrcanus* (Table 2).

All isolated strains and RT-PCR positive samples were shown to belong to genotype 1 of WNV, as determined by partial nucleotide sequencing of the E gene. Phylogenetic data of sequenced strains are presented on Fig. 1.

Serologic examination of domestic animals showed that all these species could be infected by WNV, more often in the Volga delta than in the Volga-Akhtuba area. Indices of infection rate were highest among horses, moderate among cattle and camels, and lowest among sheep (Table 3).

Discussion

Local complexes of arbovirus foci, including at least WNV, CCHFV, and DHOV, all recognized human pathogens, had been described in southern Russia [18–22]. In the course of laboratory manipulation with DHOV five cases of illness occurred with abrupt onset, marked general toxicity symptoms of subcortical pyramidal tract involvement and other lesions of the central nervous system [18]. Natural CCHF and DHOV foci in the world, as in our country, essentially coincide with the area of the main virus host and vector, *H. marginatum* ticks. In southern Russia this is the landscape belt of dry steppes, semi-deserts and deserts. In northern Russia the occurrence of these viruses is limited by isotherm of the sum of effective temperatures 3000°C [18,21,22]. Recently, a number of cir-

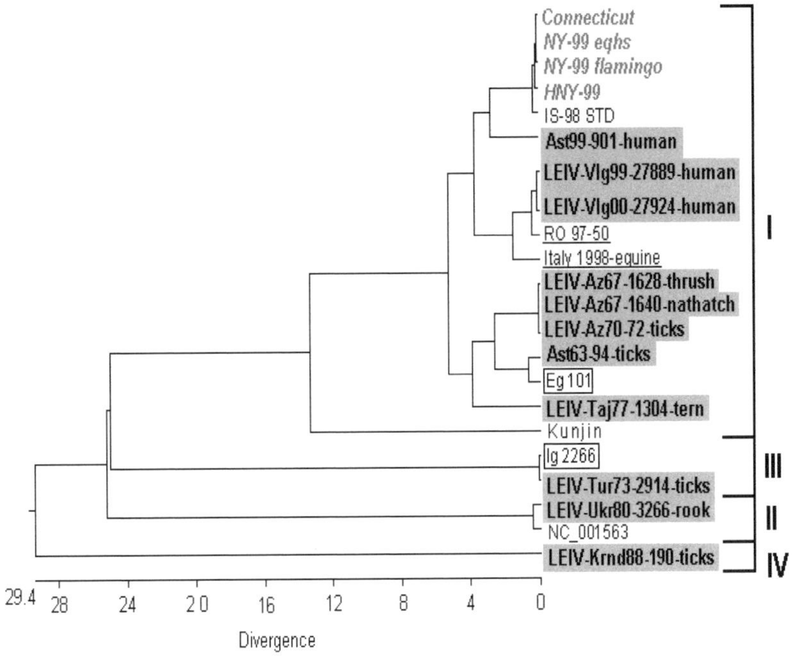

Figure 1. Phylogenetic tree for historical and epidemic West Nile virus strains. Roman numbers designate genetic lineage. Font style designations for West Nile virus strains: **historic from Soviet Union**; **Russian epidemic strain**; Topotype; European epidemic strain; ***American epidemic strain***.

cumstances have led to a considerable increase in ticks. This produced the emergence of epidemic situation for CCHF [17]. Comparative studies were done with 310–529 fragments of S-segment genomes of epidemic strains isolated during the last four years from patients and ticks and strains isolated earlier in the south of Russia and Central Asia, as well as the strains from South Asia and Africa. These showed that in southern Russia, strains were genetically similar to those circulating 30 years previously, and that nucleotide sequence differences did not exceed 1–2%. Thus, the molecular evolution of CCHF virus is progressing very slowly [17,36]. *H. marginatum* ticks are important for the long-term persistence of virus [20], for instance during winter in anthropogenic foci of WNV in the Volga delta region.

We also isolated WNV, CCHF and DHOV from crow and hare and pre-imago stages of *H. marginatum* ticks collected from them [25], and developed diagnostic tests to detect evidence of human infection with these viruses.

In the USA, WNV was detected by RT-PCR in New York State and elsewhere in 2000. Virus was detected among dead birds of 162 species, with highest indices in American crows (*Corvus brachyrhynchos*), which had 67 % mortality in the epidemic epicenter; dead birds were found two weeks before the first human illness occurred [2,12,26]. ELISA for detection of antibodies to WNV was also used for the New York studies. Antibodies were found in birds of 23 species [8]. In studies done 30–40 years ago in the Volga delta region WNV was isolated from glossy ibis, bittern, blackbird, nuthatch, herring gull, and crow [18–20]. Our data showed that the greatest involvement in WNV circulation was among cormorants, twice the prevalence in coots, and five times that of herons, gulls, and terns in natural habitats. Positive results in anthropogenic biocenosis were found among ground-feeding birds, especially among corvids. By RT-PCR, we determined that the most important mosquito vectors of WNV in anthropogenic biocenosis are the ornithophilic species *An. hyrcanus, Cx. pipiens, Cx. modestus* and the cattle-feeding species *An. messae*. In natural habitats ornithophilic species *An. hyrcanus* in the middle belt and *Coq. richiardii* and *An. hyrcanus* in low Volga delta region are most important. Overall, the infection rate of mosquitoes in anthropogenic biocenosis was twice that of the rate in natural habitats.

During the 1996 epidemic outbreak of WNV in Romania the principal vector was *Cx. pipiens* [32]. In the USA WNV has been isolated from *Cx. pipiens, Cx. restuans, Cx. salinarius, Culiseta melanura* [1,2] and from *Ae. vexans, Ochlerotatus triseriatus, Psorophora ferox,* and *An. punctipennis*. It appears likely that *Cx. pipiens* and *Cx. restuans* are the primary enzootic and epizootic vectors among birds, *Cx. salinarius* for humans, and *Ae. (Ochlerotatus) spp.* for horses [12,35]. Under experimental conditions the mosquitoes most susceptivle to infection with WNV were *Cx. restuans* and *Cx. salinarius*; *Cx. Quinquefasciatus*, and *Cx. nigripalpus* were of moderate sensitivity and *Coquillettidia perturbans* was of low sensitive [8,31]. Our virological and serological data showed a high level of WNV activity in the Volga delta region, as compared with the Volga-Akhtuba ecosystem. Serologic investigation of domestic animals were useful as indicators of WNV activity.

Phylogenetic analysis of full-dimension genomes demonstrated the following: Astrakhan epidemic strains were similar to epidemic strains from New York and Israel (nucleotide differences 2.7%); Volgograd strains were similar to epidemic strains from Romania (nucleotide differences 0.4%). It is likely that the New York outbreak originated as a result of virus introduction by infected mosquitoes from the Mediterranean by air and it is possible that the virus was introduced from Black Sea ports, by infected mosquitoes within holds of ships. Recent epidemic strains are genetically different from strains isolated in Russia in the same region from various natural sources 20–30 years ago, during which interval there was an absence of epidemics (nucleotide differences about 6%). In addition to strains of genotype 1 WNV, we isolated a strain from a corvid (*Corvus frugilegus*) belonging to genotype 2, a strain from *Hyalomma detritum* ticks belonging to the Indian genetic lineage 3 (nucleotide distance of which from the first and second groups is about 25%), and the strain from *Dermocentor marginatus* ticks that is quite different from all recognized isolates (nucleotide differences from all three groups about 30%). Using 20% nucleotide differences as a criterion, we are able to differentiate four genetic lineages of WNV, one of which, is known thus far only from Russia [17,28,29]. It is of practical importance that the epidemic situation of the last three years probably was based not only by natural-social factors but also by alterations in the genetic characteristics of circulating viral populations. On the basis of these molecular genetic studies of epidemic and historic WNV strains, we developed ELISA-based systems to detect virus antigens and, as well, developed specific IgM antibody and IgG antibody assays, as well as PCR assays that are effective for detection of all four genotypes of WNV.

Acknowledgements

The work was partly supported by ISTC Grant 2087.

The authors thank the following for their assistance – in **Astrakhan**: Guzhvin A.P. (Governor of Astrakhan region), Osipov A.A., Mikhailov G.M., Agoshkov V.M., Grishanova A.P., Zlobina L.B., Ustaev V.M., Zimin A.V., Stopkin N.V., Binkov S.V.; **Ikryanoe**: Odolevsky E.I., Kodyakov P.N., Kodyakov S.N., Shchetinkin P.F.; **Liman**: Ibragimov R.M., Pavly V.I., Seliverstov E.I.; **Akhtubinsk**: Shatilov V.P., Finogenov O.V., Lapenko V.V., Teplinskaya L.V.; **Kharabaly**: Chekrizov P.F., Moiseeva O.V., Inozemtseva I.G.; **Krasny Yar**: Volodin V.N., Turlaev Yu.V.; **Moscow**: Vakar E.I., Ivanova M.V., Belonozhkina G.M.; **Fort Collins** (USA): Tsuchiya R.

References

[1] Andreadis TG, Anderson JF, Vossbrinck R (2001) Mosquito surveillance for West Nile virus in Connecticut, 2000: isolation from *Culex pipiens, Cx. restuans, Cx. salinarius*, and *Culiseta melanura*. Emerg Infect Dis 7: 670–674.

[2] Bernard KA, Maffei JG, Jones SA, Kauffman E, Ebel G, Dupuis A et al (2001) West Nile virus infection in birds and mosquitoes. Emerg Infect. Dis: 679–685.

[3] Berthet FX, Zeller HG, Drouet MT, Rauzier J, Digoutte JP, Deubel V (1997) Extensive nucleotide changes and deletions within the envelope glycoprotein gene of Euro-African West Nile viruses. J Gen Virol 78: 2293–2297.

[4] Briese T, Rambaut A, Pathmajeyan M, Bishara J, Weinberger M, Pitlik S, LipkinWI (2002) Phylogenetic analysis of human isolate from the 2000 Israel West Nile virus epidemic. Emerg Infect Dis 5: 528–531.

[5] Brinton MA (2002) The molecular biology of West Nile virus: A new invader of the Western Hemisphere. Annu Rev Microbiol 56: 371–402.

[6] Centers for Disease Control and Prevention. (2002) West Nile virus activity - United States, September 5–11, and Texas, January 1 – September 9, 2002, Morb Mortal Wkly Rep 51: 812–823.

[7] Ebel GD, Dupuis AP, Ngo K, Nickolas D. Kauffman E, Jones S et al (2001) Partial genetic characterization of West Nile virus strains, New York state, 2000. Emerg Infect Dis 7: 650–653.

[8] Goddard LB, Roth AE, Reisen WK, Scott TW (2002) Vector competence of California mosquitoes for West Nile virus. Emerg Infect Dis 8: 1385–1391.

[9] Goodman RA, Bauman CF, Gregg MB, Videtto JF, Stroup DF, Chalmers NP (1990) Epidemiologic field investigations by the Centers for Disease Control and Epidemic Intelligence Service, 1996–1987. Public Health Reports 105: 604–610.

[10] Huang C, Slater B, Rudd R, Parchuri N, Hull R, Dupuis M, Hindenburg A (2002) First isolation of West Nile virus from a patient with encephalitis in the United States. Emerg Infect Dis 8: 1367–1371.

[11] Koller, M., Gresikova, M., Berencsi, Gy., Schablik, M.: Hemagglutination inhibition antibodies to arbovirues in the population of Hajdú-Bihar district, Hungary. Folia Parasitol. 16, 75–79, 1969.

[12] Kulasekera VL, Kramer L, Nasci RS, Mostashari F, Cherry B, Trock SC, Glaser C, Miller JR (2001) West Nile virus infection in mosquitoes, birds, horses, and humans, Staten Island, New York, 2000 (2001). Emerg Infect Dis 7: 722–725.

[13] Lanciotti RS, Ebel GD, Deubel V, Kerst AJ, Murri S, Meyer R et al (2002) Complete genome sequences and phylogenetic analysis of West Nile virus strains isolated from the United States, Europe, and the Middle East. Virology 298: 96–105.

[14] Lanciotti RS, Roehrig JT, Deubel V, Smith J, Parker M, Steele K et al. (1999) Origin of the West Nile virus responsible for an outbreak of encephalitis in the northeastern United States. Science 286: 2333–2337.

[15] Langmuir AD (1980) The Epidemic Intelligence Service of the Center for Disease Control. Public Health Reports Sept.–Oct. 95: 470–477.

[16] Larichev VF, Butenko AM, Rusakova NV, Shishkina LV, Zhukov AN, Shishkina EO, Lvov DK (2002) Indexes of specific antibodies to West Nile virus among reconvalescents. Vopr Virusol 47: 11–13 (In Russian).

[17] Lvov DK (2002) Importance of Emerging-ReEmerging Infections for biosecurity. Vopr Virusol 47: 4–7 (In Russian).

[18] Lvov DK (1994) Arboviral Zoonoses on Northern Eurasia (Eastern Europe and the Commonwealth of Independent States). In: Beran GW, Steel JH, editors. Handbook of zoonoses 2nd ed Boca Raton: CRC Press: 237–260.

[19] Lvov DK (2000) West Nile fever. Vopr.Virusol 45: 4–9 (In Russian).

[20] Lvov DK, Ilichev VD (1979) Migrations of birds and transfer of infections agents. "Science" Moscow: 1–270 (In Russian).

[21] Lvov DK, Klimenko SM, Gaidamovich SY (1989) Arboviruses and arboviral infections. "Meditsina" Moscow: 1–335 (In Russian).

[22] Lvov DK, Deryabin PG, Aristova VA, Butenko AM, Galkina IV, GromashevskyVL, Davydova AA, Kolobukhina LV, Lvov SD, Scchelkanov MY (2001) Atlas of distribution of zoonotic viral infections on territory of Russia. Ministry of Public Health of Russian Federation, Moscow, 193 pages (In Russian).

[23] Lvov DK, Butenko AM, Gromashevsky VL. Larichev VP, Salko VN, Kovtunov AI, Galimzyanov KM, Platonov AE, Morozova TN, Khutoretskaya NV, Shishkina EO, Skvortsova TM (2000) Isolation of two strains of West Nile virus during an outbreak in Southern Russia, 1999. Emerg Infect Dis 6: 373–376.

[24] Lvov DK, Butenko AM, Gaidamovich SY, Larichev VF, Leschinskaya EV, Lazarenko VV et al (2000) Epidemic outbreak of meningitis and meningo-encephalitis in Krasnodar territory and Volgograd region provoked by West Nile virus. Vopr Virusol 45: 37–38 (In Russian).

[25] Lvov DN, Jarkenov AF, Aristova VA, Kovtinov AT, Gromashevsky VL, Galkina IV et al (2002) Isolation of Dhori (*Orthomyxoviridae*, *Thogotovirus*) and Crimean-Congo hemorrhagic fever (*Bunyaviridae, Nairovirus*) from hare (*Lepus europaeus*) and collected from ticks *Hyalomma marginatum* in the middle part of Volga-delta, Astrakhan region, 2001. (2002) Vopr Virusol 47: 32–36 (In Russian).

[26] Marfin AA, Petersen LR, Eidson M, Miller J, Hadler J, Farello C et al (2001) Widespread West Nile activity, Eastern United States, 2000. Emerg Infect Dis 7: 730–735.

[27] Nasci RS, White DJ, Stirling H, Oliver J-A,Daniels TJ, Falco RC et al (2001) West Nile virus isolates from mosquitoes in New York and New Jersey, 1999. Emerg Infect Dis 7: 626–630.

[28] Prilipov AG, Kinney R, Samokhvalov EI, Savage HM, Alkhovsky SV, Tsychia R, Gromashevsky VL, Sadykova GK, Shatalov AG, Vyshemirsky OI, Usachev EV, Mokhonov VV, Voronina AG, Butenko AM, Larichev VF, Gubler DJ, Lvov DK (2002) Phylogenetic analysis of a new variants of West Nile virus. Vopr Virusol 47: 36–41 (In Russian).

[29] Prilipov AG, Samokhvalov EI, Lvov DK, Gromashevsky VL, Butenko AM, Larichev VF et al (2001) Genetic analysis of West Nile viruses isolated on the south of Russian Plain (Volgograd and Astrakhan regions) in 1999. Vopr Virusol 46: 8–12 (In Russian).

[30] Scherret JH, Poidinger M, Mackenzie JS, Broom AK, Deubel V et al (2001) The relationships between West Nile and Kunjin viruses. Emerg Infect Dis 7: 697–705.

[31] Sardelis MR, Turell MJ, Dohm DJ, O'Guinn ML (2001) Vector competence of selected North American *Culex* and *Coquilletidia* mosquitoes for West Nile virus. Emerg Infect Dis 7: 1018–1022.

[32] Savage HM, Ceianu C, Nicolescu G, Karabatsos N, Lanciotti R, Vladimiresku A et al. (1999) Entomologic and avian investigations of an epidemic of West Nile fever in Romania in 1996, with serological and molecular characterization of a virus isolated from mosquitoes. Am J Trop Med Hyg 61: 600–611.

[33] Shishkina EO, Butenko AM, Gaidamovich SY, Larichev VF, Zhuravlev VI, Shcherbakova SA et al. (2002) Immune structure of population to West Nile virus in Astrakhan region in 1998 and 1999. Vopr Virusol 47: 13–17 (In Russian).

[34] Virus Taxonomy; 7th Report of the ICTV (2000) Eds: Van Regenmortel MHV, Fauquet CM, Bishop DHL, Carstens EB, Estes MK, Lemon SM, Maniloff J, Mayo MA, McGeoch DJ, Pringle CR, Wicner RB Academic Press San Diego, California. 1162 pages.

[35] White DJ, Kramer LD, Backenson PB, Lukacik G, Johnson G, Oliver JA et al (2001) Mosquito surveillance and polymerase chain reaction detection of West Nile virus. New York state. Emerg Infect Dis 7: 643–649.

[36] Yashina L, Petrov VC, Vyshemirsky OI, Aristova VA, Moskvina TM, Lvov DK et al (2002) Characteristics of Crimean-Congo hemorrhagic fever virus circulating in the Russia and Central Asian Republics. Vopr Virusol 47: 11–15 (In Russian).

Emerging Biological Threat
G. Berencsi et al. (Eds.)
IOS Press, 2005

Natural Foci of Classical and Emerging Viral Zoonoses in Hungary

Emőke FERENCZI[a], Gábor RÁCZ[b], Gábor FALUDI[c], Alíz CZEGLÉDI[a], Ilona MEZEY[a] and György BERENCSI[a]
[a] *"B. Johan" National Center for Epidemiology, Division of Virology, Budapest, Hungary*
[b] *University of New Mexico. Department of Biology, Albuquerque, NM*
[c] *Medical Research Institute of the Hungarian Defence Forces, Budapest, Hungary*

The most important duty of a clinical virologist working in a public health laboratory is to find out the aethiological diagnosis of infectious diseases, especially of outbreaks of emerging/re-emerging diseases. In the case of zoonotic diseases search for natural foci and surveillance constitute the basis of the work that complete this requirement. The same preparedness is indispensable for the recognition of an accident involving biological and toxin warfare (BTW) agents [1].

1. Introduction

During the past several decades it turned out, that the two most important viral zoonoses in Hungary are the tick-borne encephalitis and the haemorrhagic fever with renal syndrome/nephropathia epidemica. Most recently West Nile virus infections emerged, hopefully it will not be established as the third endemic viral zoonosis in Hungary.

2. History

The most important events and stages of the progress of viral zoonosis research and diagnosis is summarized in Table 1.

2.1. Systematic Field Surveys

Ticks, mosquitoes and small mammals have been collected in different regions of the country. Over 60 TBE, 2 West Nile, 2 Crimean-Congo haemorrhagic fever, and several other arbovirus strains were isolated. TBE virus infected ticks and small mammals were found everywhere in Hungary, but mostly in the woodlands of Transdanubia and of the Northern part of the Intermediate Chain of Mountains [2]. The local serosurvey of the population revealed the presence of West Nile virus specific and Bunyavirus specific antibodies in rural population [3]. Hungary is situated in the Carpathian basin, the map (Fig. 1.) shows the different geographical regions of the country and the distribution of natural foci of tick-borne encephalitis virus.

The field surveys for hantavirus natural foci showed that the seroprevalence for human pathogen hantaviruses among rodents is about 7.25 per cent. Molecular analysis of viral

Table 1. Summary of the milestones in search of zoonotic viruses in Hungary.

1952	First isolation of Tick-Borne Encephalitis Virus (TBEV) (Fornosi, F., Molnár, E.)
1953–54	Epidemiological and clinical recognition of endemic haemorrhagic fever with renal syndrome (HFRS) [6]
1958	Regular serodiagnosis of TBE (methods: HI, occasionally CF, and NT)
1966–85	Systematic field survey of Arboviruses
1977	Vaccination against TBE - endangered people, free of charge.
1980	Rapid serodiagnosis of TBE (methods: IFA, chromatographic separation of sera before IgM determination) [5]
1987	Regular serodiagnosis of HFRS (methods: IFAT, High density particle agglutination and ELISA) [4]
1991	TBE vaccine put at the disposal for everyone in the pharmacy
1997–2000	Systematic field survey of hantaviruses
1998	First isolation of hantaviruses in Hungary
1998 and 2000	Isolation and sequencing of four Hungarian hantaviruses
2003	Serological diagnosis of the first indigenous West Nile virus infections with CNS complications

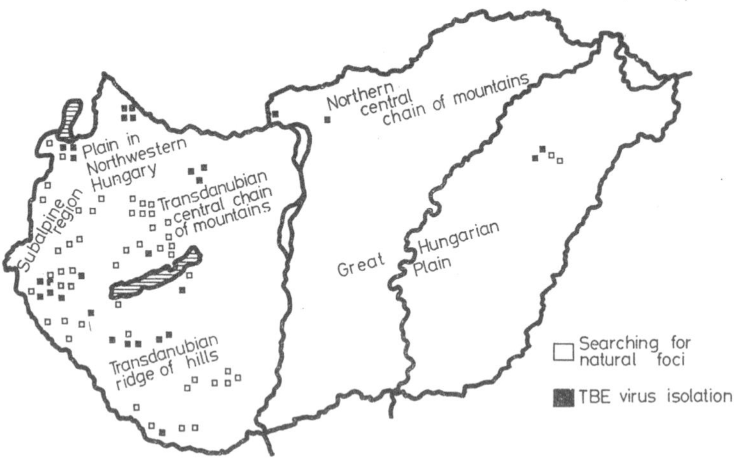

Figure 1. Geographical regions of Hungary, and natural foci of tick-borne encephalitis virus. (Courtesy of E. Molnár).

nucleic acid isolates from organs of four rodents proved directly the presence of viruses belonging to Puumala and Dobrava/Belgrade species in Hungary. The Dobrava/Belgrade type viruses were found in two different rodent species, *id est Apodemus flavicollis* and *Apodemus agrarius* [4].

2.2. Serodiagnostic Verification of Diseases Caused by Tick-Borne Encephalitis Virus

Diagnostic tests for TBE in blood sera taken from patients with suspected TBE and hantavirus infection have been performed regularly since 1958 and 1987, respectively. Over the years more and more TBE cases have been recognized and confirmed. Until 1960 a few TBE cases were recognized by clinicians and confirmed in the virological laboratory, while several outbreaks of Haemorrhagic Fever with Renal Syndrome (HFRS) have been observed in military camps. These illnesses could be confirmed using specific methodology almost 30 years later as hantavirus infections emerged [5]. A few sporadic illnesses have

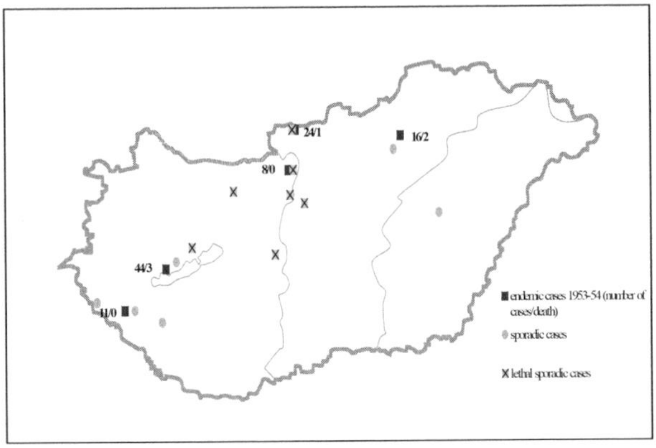

Figure 2. Endemic and sporadic HFRS cases in Hungary (1952–1961).

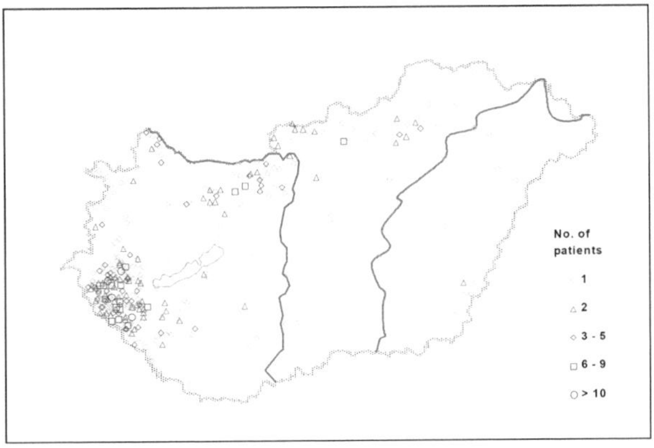

Figure 3. Serologically verified tick-borne encephalitis cases 1961–70 (660 patients).

been diagnosed among hospitalized patients, too. The geographical distribution of their homes are shown in the map in Fig. 2.

Since 1960 increased number of TBE patients have been recognized as well as confirmed. Most of them occured in the Transdanubian region, as is shown in Fig. 3 on the map.

Between 1971 and 1980 it became obvious that TBE is a serious public health problem in Hungary with hundreds of hospitalized clinical diseases every year. The geographic distribution of disease outbreaks are shown in Fig. 4.

The most patients suffering from TBEV infection were diagnosed between 1981 and 1990. The annual average was 300 illnesses/year. Geographical distribution of the TBE infections changed in this decade (Fig. 5), most likely due to better awareness in regions earlier thought to be less endemic, and to the large number of people vaccinated in the formerly most endemic regions.

2.2.1. Obvious Decrease of TBE Infections from 1991 in Hungary

Between 1991 and 2000 the number of diagnosed TBE cases has significantly decreased, while the number of recognized and serologically confirmed hantavirus infections increased. Identified sites of infection of TBE patients are shown in Fig. 6.

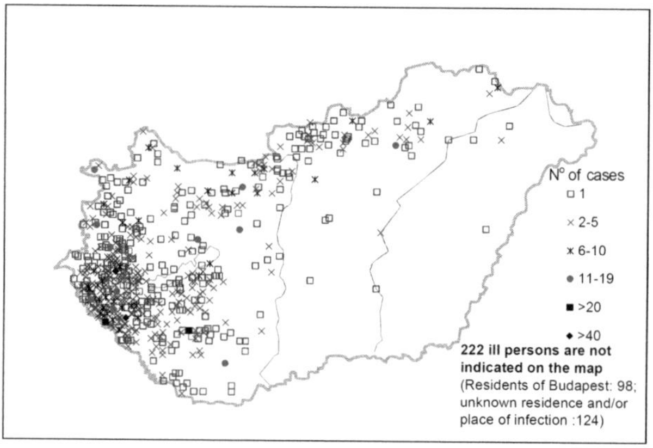

Figure 4. Serologically verified tick-borne encephalitis infections, between 1971–80 (1894 patients).

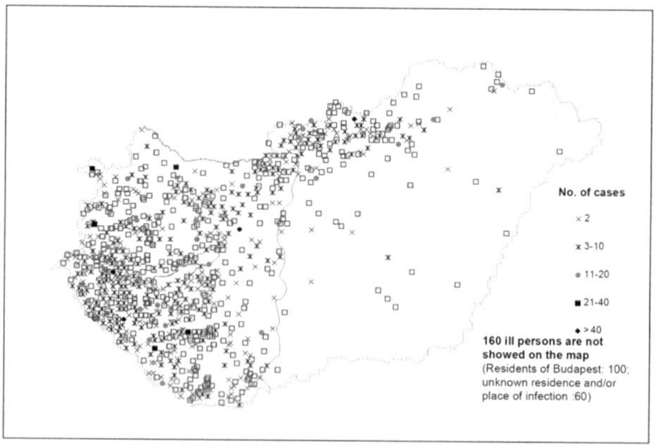

Figure 5. Serolocically proven tick-borne encephalitis infections, between 1981–90 (2878 patients).

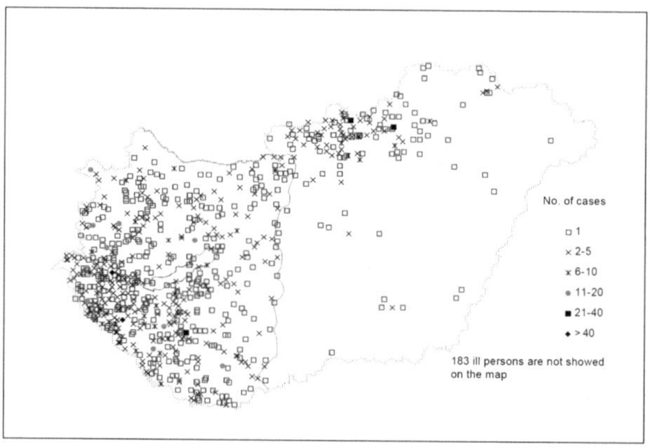

Figure 6. Serologically proven tick-borne encephalitis infections between 1991–2000 (1874 patients).

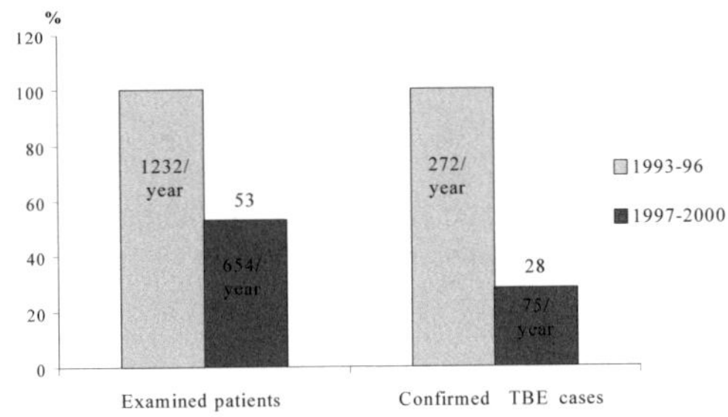

Figure 7. Comparison the standardised mean annual numbers of examined patients (100 %) and those of laboratory confirmed TBE cases during the two periods of different funding.

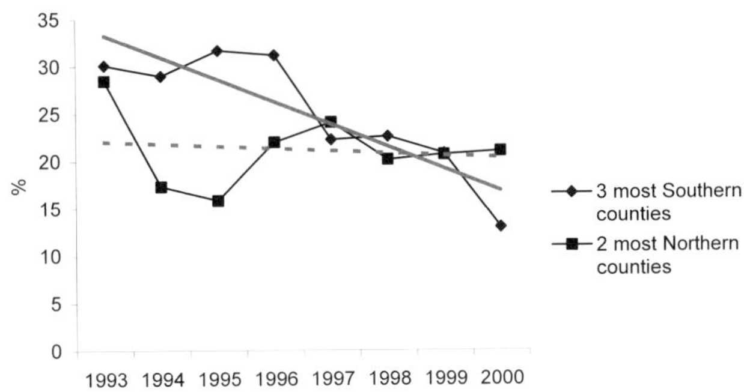

Figure 8. Proportion of the number of confirmed TBE patients of those examined (*Solid line*: linear regression in the Southern region, *Broken line*: regression in the Northern region of Hungary).

The decrease was not a gradual one, since 1997 the annual number of patients has decreased dramatically. Concomitantly, also the funding system of the health services has changed. In other words, since 1997 the hospitals had to pay for the specific serological diagnostics. Previously the expenses of specific serological diagnostics had been included into the budget put at the disposal of the hospitals for the treatment of different clinical illnesses. The diagram shown in Fig. 7 indicates, however, that in spite of the reduced absolute number of the TBE-specific tests performed, the proportion of the verified specific diseases decreased significantly. In spite of the fact, that the annual number of examined patients diminished from 100% to 53% of the previous four years, the proportion of confirmed cases dropped from 100% to 28% (i.e. 3.5 times) of that confirmed during the preceding years.

The percentage of diagnostic hits by year is shown in Fig. 8.

The differences are significant. Comparison of the tendency shows differences between 3 Southern endemic counties and two Northern endemic counties. Decline of the trend-line was found in the case of Southern region, but it remained at the same level in the Northern region.

The decrease of TBE diseases could have been caused by different reasons or by the combined effects of them:

1) due to increased level of vaccination history of the population in endemic regions,
2) ecological changes (i.e. increase of the mean temperature),

Figure 9. Geographical distribution of patients with laboratory confirmed hantavirus infections (1998–2000) (modified from reference No. 4).

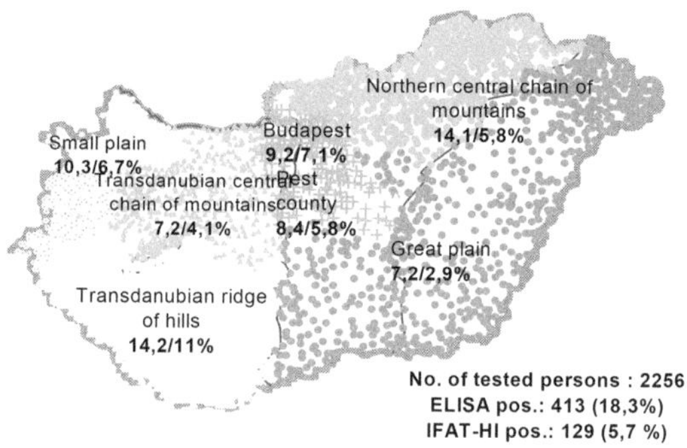

Figure 10. TBEV seropositivity in the geographical regions of Hungary on the basis of the seroepidemiological survey in 1999–2000.

3) due to social – economic changes, including the modified financing.

The observed data raise the possibility, that the climate change might influence the incidence of the disease in Hungary, because our country is located near to the southern border of the natural habitat of *Ixodes ricinus*.

The increasing number of diagnosed hantavirus infections could be the result of better awareness, even if there are some counties without sampling for hantavirus laboratory tests. Fig. 9 shows the diagnosed hantavirus caused illnesses between 1998 and 2000.

A national serosurvey was conducted in 2000, using sera of healthy individuals over 20 years, collected in Hungary in 1999. 2256 sera were tested for presence of TBEV and hantavirus specific antibodies. The results indicated, that 5.72% and about 10% of healthy persons can be considered to be seropositive for TBEV and hantaviruses, respectively. TBE and hantavirus seropositive persons were found to be present in all areas of Hungary, nevertheless the majority of the TBE seropositive inhabitants are concentrated in 5 forested counties.

The map shown in Figure 10 indicates the seroprevalence of TBE virus-specific antibodies in different geographical territories. Data of the capital (Budapest) and Pest county are indicated separately, because they belong to 3 different regions.

Although it is known, that only minimal foci were shown to be populated by infected ticks in the Great Plain. The voluntary immunization practice has resulted in a rather high percentage of seropositive population. Their proportion, however, was significantly lower, than that in the Transdanubian Ridge of Hills (14.2/11%).

Finally one has to mention, that during August, 2003 emerged the first case of indigenuous West Nile virus infection in Hungary. Up to the completion of this manuscript 14 sporadic cases with CNS symptoms have been confirmed. Residences of these patients are in the geographic regions that are non-endemic for TBEV.

The future will indicate whether the climate change was responsible for these modifications, or one has observed a merely short-term fluctuation in the natural foci. The occurrence of West-Nile virus was shown earlier, probably due to the movement and appearance of wild geese (and other migratory birds) in the Hortobagy National Park of the Great Hungarian Plain [3]. The continuous surveillance might reveal the role of local or global meteorological relations.

References

[1] Berencsi G. and Faludi G. (2001) The role of biotechnology in protection against BTW agents – Overview from a medical point of view and identification of the symptoms, in A. Kelle et al. (eds.), The Role of Biotechnology in Countering BTW Agents, 215–225. Kluwer Academic Publishers.

[2] Molnár E. (1982): Occurrence of tick-borne encephalitis and other arboviruses in Hungary, in Geographia Medica 12, pp. 78–120.

[3] Koller M., Gresikova M., Berencsi Gy., Schablik M.: Hemagglutination inhibition antibodies to arbovirues in the population of Hajdú-Bihar district, Hungary. Folia Parasitol. 16, 75–79, 1969.

[4] Ferenczi E., Rácz G., Szekeres J., Balog K., Tóth E., Takács M., Csire M., Mezey I., Berencsi Gy., Faludi G.: Újabb adatok a hazai hantavírusok népegészségügyi jelentőségének vizsgálatához. (New data contributing to the public health importance of local hantaviruses – in Hungarian) Orv Hetil. 2003 Mar 9; 144(10):467–74.

[5] Takenaka A., C.J. Gibbs, and D.C. Gajdusek (1985) Antiviral Neutralizing Antibody to Hantaan Virus as Determined by Plaque Reduction Technique, Archives of Virology 84:197–206.

[6] Trencséni T., Keleti B. (1971): Clinical aspects and epidemiology of Haemorrhagic fever with renal syndrome. Analysis of the clinical and epidemiological experience in Hungary. Akadémiai Kiadó, Budapest.

Emerging Biological Threat
G. Berencsi et al. (Eds.)
IOS Press, 2005

Coronavirus Causing Severe Acute Respiratory Syndrome (SARS), a New Highly Infectious Emerging Threat

Younes Ali SALEH*, Péter GYARMATI, Zoltán KIS and György BERENCSI
"B. Johan" National Center for Epidemiology, Division of Virology, Gyáli str. 2-6, Budapest, Hungary

Veterinary medicine had many practical problems with coronaviruses discovered in the thirties. Molecular biologists had to confront new problems when the unique size and replication strategy of the virus had been recognised first. The majority of the medical society became aware on the importance of the Coronavirus Family only upon the emergence of the SARS-CoV.

1. History of the SARS-CoV Epidemic

On 27th November 2002 WHO has received an unofficial message, that severe influenza like illnesses occurred in southern regions of the continental China. Only on the 11th of February, 2003 information has been obtained by the WHO Office in Geneva, that the infection rate and mortality was high among health professionals. The official report was obtained on the 14th of February.

The father of a visiting family died on the 19th of February. Influenza A/H5N1 has been detected in the samples taken from him.

The disease has been imported to Vietnam on the 26th of February, therefore, WHO delegated a medical team on the 5th of March in order to support the clarification of the etiology. Carlo Urbani, M.D. [1] was a member of this group, who acquired the infection. He has realized, that it is a new disease. Many publications have been dedicated to his memory, since he deceased thereafter [1–4].

The denomination "SARS" has been accepted on the 15th of March (Klaus Stőhr, personal communication). The case definition has been summarized, and circulated in early April in the majority of member states. The cooperation of a network of 11 reference laboratories has been initiated by the WHO on the 17th of March, which were located in 9 countries. Their efforts made it possible the isolation and characterization of a new coronavirus as the etiologic agent by the 16th of April [3].

The innocuous reagents enabling the performance of rapid diagnostic tests have been developed in Germany following the transmission of the disease into the country [2,5]. The control samples have been distributed free of charge for the National Laboratories. First time in the history of WHO travel-restrictions have been recommended at first on the 27th of March, then repeatedly on the 2nd and 23rd April in order to restrict the number of travelers with destination to the countries where this epidemic happened.

The first epidemiological examinations concerning the behavior of the new SARS Coronavirus (SARS-CoV) have been completed in Vietnam. They could stop the appearance of new illnesses by the 28th of April introducing isolation of the patients, caranteen of contact persons, and by the education of health personnel to wear the provided safety garment appropriately.

It was suggested, that symptom-less people also might shed the virus with secretions therefore strict travel restrictions can be released. Only the departure of people with fever was prevented thereafter only from countries or cities, where SARS patients have been treated and new infections have been registered within the last 3 weeks.

2. Problems of Epidemiology and Diagnostics in Connection with the SARS Epidemic

The spread of the disease has been followed up first in Hong-Kong [3,6,7], in Canada [8] and in Singapore [1] according to the first publications. The analysis of the cases in the Filippine Islands was published only later (8th June, 2003).

2.1. Onset of the Clinical Symptoms After Infection [4,9]

The first sign of the disease was fever above 38.5°C followed by influenza like symptoms (dizziness, muscle pain, loss of appetite, sore throat, giddiness, shivers, convulsions and diarrhea. The onset of clinical symptoms was 2 to 10 days following infection.

Dyspnoe has developed in one third of the patients [6] and according to the first publication from Hong Kong which has proven convincingly the etiologic role of the Coronavirus the onset of dyspnoe occurred several days after the onset of the clinical symptoms [6]. The publication has documented, that the earlier initiation of treatment has reduced significantly the frequency of the lethal outcome of the disease. The beneficial effect was not due to the effectiveness of the drugs against the SARS-CoV, but due to the inhibition of the replication of bacteria coinfecting the patients simultaneously modulating the harmful immunologic defence mechanisms activated by the infections. Associated microbial infections aggravate the clinical outcome of the disease. Such effect has been documented in connection with infections caused by metapneumovirus, influenza viruses and several bacteria (*C. pneumoniae, M. pneumoniae és S. pneumoniae)* [8].

The risk of the infection of other people was lower in earlier patients which were admitted to the specialized hospital departments [6]. In April the majority of the newly recognized suspected patients have been isolated in wards in contrast to the proportion admitted to hospitals in the first half of the epidemic. In Hong-Kong 910 patients have been admitted to the hospital within 3 days after the onset of the symptoms, in contrast to 20 patients who could be isolated only more than 10 days after the onset of the clinical disease.

The majority of the patients have recovered and released from the hospital after 3 weeks. In some cases, however, recovery after 40–50 days was also seen. Among the patients above 60 years of age, the occurrence of 60 days hospitalization was frequent and the lethal outcome was about 50 % in this age group.

2.2. The Follow Up of the Transmission and Spread of the Infection Has Been Made Possible by the Newly Developed Specific Molecular Diagnostic Reagents

Molecular examinations of isolated virus strains including nucleotide sequence analysis of different viral genes (using comparative analysis of "unrooted evolutionary trees") indicated that the epidemic spread independently, following different epidemiological chains. The authors succeeded to separate the viruses of the first wave (Beijing, Quang-Dong and Hong-Kong) from those of viruses of the "second wave" (Singapore, Canada, Hong-Kong and Hanoi). In the meantime the disease has reappeared in 2004 in China. These virus iso-

Table 1. Standardized number of SARS in different cohorts of Hong-Kong [6].

Age of patients (years)	Number of patients per 10,000 inhabitants
0–19	0.8
20–34	2.8
35–59	2.3
Above 55	Increasing up to 3.2

lates were found to be more closely related to the virus strains obtained from cibet cats indicating the variability and diversity of coronaviruses [11].

The SARS coronavirus epidemic was successfully stopped by the application of epidemiological interventions, since the viruses have been secreted by the patients only after the onset of clinical symptoms which was fever. In addition to this the transmission required close contact of the patient and the target persons (similar to the RSV infection). SARS-CoV has been transmitted mainly in the form of respiratory infection to family members and to the health care personnel. The possibilities of oral infection, and "dirty hand" transmission, however, could not be excluded.

On the basis of model experiments SARS-CoV is retaining infectivity for 3 weeks at +4 °C, for 1 or 2 days at room temperature. It has been shown that the stool of the patients carry very high loads of viral RNA-containing particles, too [3]. The fecal samples are the best diagnostic sources for the detection of the infection even after 10 days upon the onset of the disease. Fortunately all conventional disinfectants are able to inactivate the virus within minutes [1,20].

The attack rate among children and teenagers was much lower than that in the older cohorts. One might suppose, that the primary infections with classical human coronaviruses have resulted in some cross-protection against SARS-CoV which disappeared in the older cohorts.

2.3. Treatment of Suspected SARS Patients is Associated with Different Problems of Labor Health [1,20]

1) Patients have to be isolated (under low air pressure, if possible). In case a larger number of patients had to be isolated they had been located on the top floor of the hospital, and artificial air-flow have been created to the direction of the open windows [17] thus reducing the risk of transmission.
2) Patients have to wear mouth protection masks during medical examinations in order to reduce the risk of infection.
3) The health care personnel have to wear protective gloves, coats, glasses and masks at least of 95 to 100 protective capacity.

The number of health care personnel has to be reduced to the lowest possible number. Visitors should be restricted to the lowest possible number and equipped with the protective garment available for the health care personnel.

At the end of the work health personnel has to disinfect carefully hands upon the removal of the protective garment.

Following the recovery of the patients the wards or flats have to be carefully disinfected.

4) Suspected SARS patients are sampled, respiratory tampons have to be used and the samples have to be transported in transport medium in order to make detection of conventional respiratory pathogens causing atypical pneumonia (adenovirus, Chlamydia, influenza, Mycoplasma, parainfluenza, RSV) possible.

Computer tomography may facilitate the diagnosis. In the case of diseases of lethal outcome serial frozen histological sections have to be prepared from the pathological tissues in order to facilitate rapid etiological diagnosis.

3. Histopathological Changes Caused by and Cultivation of SARS-CoV

The detailed examination of the autopsy samples taken from the first 9 patients has revealed severe alveolar damage. Pneumocytes have been released from the alveolar wall and accumulated within the lumen together with erythrocytes. Swollen cells were also seen in the alveolar vesicles and between them in the interstitium. Signs of inflammation, and accumulation of macrophages and precipitated proteins could be documented. Necrotic spots and accumulation of flat endothelial cells were observed on the mucous membranes of the bronchioli. The results of the electron microscopic examination were surprising. Virus particles or inclusion bodies indicated the sites of virus replication were absent. Virus particles were only found in the cells of the swollen secretory glands of the epithelium (1, 2, 12). These results suggest, that the lethal outcome has not been caused by the extensive virus replication directly, but probably by the induced overproduction of inflammatory substances and hyperactive defence reaction.

SARS coronavirus can be easily cultivated in both monkey and human tissue culture cells [2,6,19]. Syncytia are induced by the virus infection, but no cytoplasmic inclusion bodies can be detected.

4. Cultivation and Molecular Diagnosis of SARS Virus Infection and Its Immunomodulating Effect

Rapid diagnosis could be achieved only using the molecular diagnostic techniques developed, and by the direct electron-microscopic examination of the cells obtained from bronchoalveolar lavage (BAL) samples [2,5].

The cooperation of the Bernard Nocht Institute in Hamburg, the Virology Unit of the Hong-Kong Governmental Laboratory and the Centers for Disease Control (4) developed molecular amplification primers both for nested PCR and for real time PCR based on the TaquMan chemistry [2,3,5,14]. The presence of the virus can be detected within 24 hours after the onset of the disease. Inactivated control preparations have been distributed early in May, and an external quality control test have been offered for national laboratories with the participation of 55 partners [21].

There are several procedures available for the detection of the antibodies induced by the SARS-CoV infection [2,3,5]. The most reliable results have been published first on the basis of the examination of 19 patients with blood samples. The appearance of the virus specific antibodies were observed 2 to 3 weeks after the onset of the disease. Therefore, serology cannot be used for the rapid diagnosis of the disease. The late production of antibodies indicated, that the virus has a yet uncharacterized inhibitory effect to the immune system.

5. Taxonomical Position of SARS-CoV in the Family of Coronaviridae

Coronaviruses have been described first in 1931. Veterinary medicine had to cope with severe animal epidemics of significant economical losses. Live and killed vaccines have been produced. Coronaviruses could be grouped in 3 lower taxa. SARS CoV was shown to possess similar epitopes to the transmissible gastroenteritis virus of pigs, but the sequence

Table 2. Coronaviruses and the most important illnesses caused by them.

Virus	Sub-group	Host species	Clinical illness	Preventive measure
HcoV-229E	I	Human	Upper respiratory illness	–
Feline infectious peritonitis (FIPV)	I	Cat	Infectious peritonitis	–
Transmissible gastroenteritis virus (TGEV)	I	Pig	Diarrhea, diabetes, encephalitis	Live vaccine
Porcine infectious diarrhea	I	Pig	Epidemic diarrhea	–
Canine corona-virus (CCo-V)	I	Dog	Epidemic diarrhea	Killed vaccine
HcoV-OC43	II	Human	Upper respiratory infection	–
Mouse hepatitis virus (MHV)	II	Mouse	Hepatitis and neurological illnesses	–
Bovine gastroenteritis	II	Bovine	Epidemic diarrhea	–
Porcine hemagglutinating encephalomyelitis	II	Pig	Encephalomyelitis	–
Sialodacroadenitis of rat	II	Rat	Periorbital inflammation and lymphadenitis	–
SARS-CoV	II/III	Human	SARS	–
Infectious bronchitis virus (IBV)	III	Poultry	Infectious bronchitis, kidney and neurological impairment	Live vaccine
Turkey diarrhea virus	III	Turkey	Hemorrhagic diarrhea	–
Rabbit CoV	?	Rabbit	Diarrhea	–

analysis of the growing number of isolates has revealed, that it is related to both subgenera 2 and 3 [2–4,10]. The most important coronaviruses are summarized in Table 2.

Coronaviruses represent a unique family of positive-stranded RNA viruses with envelope. The size of the genome is about 30 kb [2,5]. The thermodynamic stability of such a large polynucleotide chain requires specific molecular mechanisms (secondary structure and protective proteins) in order to prevent fragmentation of the genomes.

In spite of the positive polarity of the genome it is not translated directly by ribosomes, but 4 to 8 mRNA molecules are required in order to produce the necessary amounts of proteins for virus replication and assembly. The structure of the viral genome is summarized in Fig. 1 [1,2,4,5,16]. Structural genes are only labeled by letters. The length of the genome was found to be between 29 and 30 kb.

The viruses are coding for 23 proteins (horizontal arrows). Proteinase (PROT), polymerase (POL), helicase (HEL), sialidase (p65, hemagglutinin esterase) were shown to be non-structural proteins. Nucleocapsid protein (N) possesses non-structural, regulatory functions, too. SARS-CoV encoded protease was shown to be inhibited by Nelfinavir [11], but it was found to be related to serine proteases (rhinoviruses).

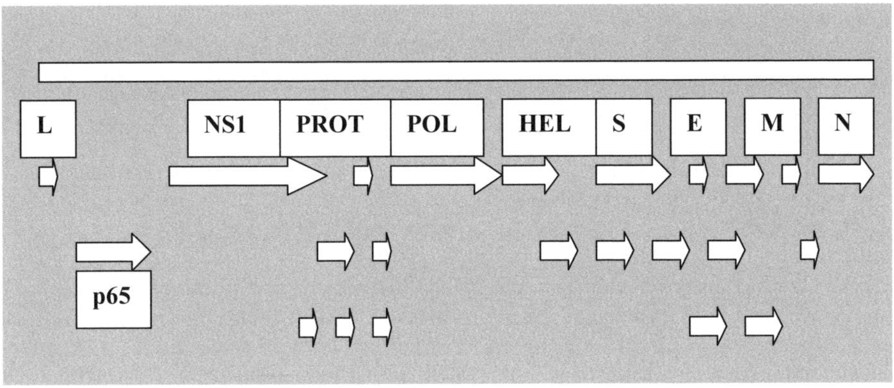

Figure 1. Genome structure of SARS-CoV. L = "leader"; NS1 = non structural protein 1; PROT = virus-coded proteinase enzyme which was found to be inhibited by Nelfinavir [11,17]; POL = polimerase; HEL = helicase, which is a metal-binding protein and possesses nukleotide triphosphatase activity; S = membrane-spike; E = envelope; M = membrane protein; N = nucleocapsid protein; p65 = protein of sialidase activity related to that of mouse hepatitis CoV [3,14,18]. Unlabelled arrows indicate ORFs NS2 to NS13. Five of them were found to be unrelated to other known coronavirus proteins [1–5].

Glycosilated proteins are incorporated into the viral envelope. "Spike" (S) protein is responsible for host-cell specificity. Interchange of domains of the spike glycoprotein by recombinant technology may modify host specificity of the virus.

5.1. The Receptor for Many Coronaviruses Was Found to Be Aminopeptidase N

The receptor for coronaviruses causing gastroenteritis was found to be Bgp2 glycoprotein. The receptor for SARS-CoV is angiotensin converting enzyme No. 2 (ACE-2). It has been suggested recently, that the expression of the ubiquitous enzyme ACE-2 on the epithelial cells of alveoli and and small intestine mucosa might play important role in the pathogenesis of the disease [23]. Antibodies to "S" protein can prevent the development of clinical disease and its pathogenetic role in illnesses characterized by demyelinization has been also suggested.

Glycoprotein "M" is also an envelope constituent required for virus adsorption and penetration. The 3rd glycoprotein of the envelope (E) has unknown function in the virus cycle, but it seems to possess pathological role (immune modulation).

5.2. Position of the Emerging Viral Zoonosis, SARS-CoV in the Evolutionary Tree of Coronaviruses

According to the unrooted evolutionary trees of 6 genes the virus has been shown to be located between the 2nd and 3rd subgenera of Coronaviruses [15] created probably by the usual recombination mechanisms of large RNA viruses [24]. According to the opinion of molecular virologists SARS-CoV is probably a "molecular mosaic". It is probably an emerging viral zoonosis, which can be controlled by epidemiological and molecular epidemiological means at the present stage of its evolution [1,2,5,11–14,24]. The possibility of its large variations in transmission and the microbiological evidences cannot be ignored [25].

Acknowledgements

The authors should like to appreciate the endavours ov veterinary virology in the clarification of natural history and pathology of Coronaviruses of veterinary interest in the rapid characterization of the disease.

The role of the WHO in last 50 years in promoting the surveillance of influenza like illnesses (ILI) and that in the rapid organization of an international co-operative SARS research has been a major contribution in the rapid identification and characterization of the virus.

The authors remember with sincere respect of Dr. Carlo Urbani, who has been victim of the medical and epidemiological research on SARS. The authors are mourning also the many hundreds of health personnel who have been infected at work due to their professional commitment. We hope their renunciation had promoted progress of the society and medicine.

Finaly the helpful advice and support of Prof. Dr. Eva Gonczol and Dr. Maria Takacs is very much appreciated.

References

[1] Ruan Y., Wei C.L., Ee L.A., et al.: Comparative full length genome sequence analysis of 14 SARS coronavirus isolates and common mutations associated with putative origins of infections. Lancet, (2003), **361**:1779–1785.
[2] Ksiazek T.G., Erdman D., Goldsmidth C.S., et al.: A novel coronavirus associated with severe acute respiratory syndrome. New England J. Med. (2003) **348**:1953–1966.
[3] Lipsitch M., Cohen T., Cooper B. et al.: Transmission dynamics and control of severe acute respiratory syndrome. Science (2003) **300**:1966–1970.
[4] Rota P.A., Oberste M.S., Monroe S.S. et al.: Characterizationof a novel coronavirus associated with severe acute respratory syndrome. Science (2003) **300**:1394–1399.
[5] Drosten C., Günther S., Preiser W. et al.: Identification of a novel coronavirus in patients with severe acute respiratory syndrome. New England J. Med. (2003) **348**:1967–1976.
[6] Donnelly C.A., Ghani A.C., Leung G.M., et al.: Epidemiological determinants of spread of causal agent of severe acute repiratory syndrome in Hong Kong. Lancet, (2003), **361**:1761–1766.
[7] Tsang K.W., Ho P.L., Ooi G.C., et al.: A cluster of cases of severe acute respiratory syndrome in Hong Kong. New England J. Med. (2003), **348**:1977–1985.
[8] Poutanen S.M., Low D.E., Henry B., et al.: Identification of severe acute respiratory syndrome in Canada. New England J. Med. (2003), **348**:1995–2005.
[9] Peiris J.S.M., Lai S.T., Poon L.L.M. et al.: Coronavirus as a possible cause of severe acute respiratory syndrome. Lancet (2003), **361**:1319–1325.
[10] Holmes K.V.: SARS-associated coronavirus New England J. Med. (2003) **348**:1948–1951.
[11] Poon L.L.M.: Detection of SARS-CoV in animals and humans. International Conference on SARS, (2004) May 8–11, Lübeck.
[12] Baric R.S., Yount B., Hensley L. et al.: Episodic evolution mediates interspecies transfer of murine coronavirus. J. Virol. (1997) **71**:1946–1955.
[13] Breslin J.J., Mork I., Smith M.K. et al.: Human coronavirus 229E: receptor binding domain and neutralization by soluble receptor at 37 degrees C. J. Virol. (2003) **77**:4435–4438.
[14] Enjuanes L., Cavanagh D.: Coronavirus. In Viruses (Eds. Tidona C.A., Darai G.), Springer, Berlin, Heidelberg, New York, 2002.
[15] Riley S., Fraser C., Donnelly C.A. et al.: Transmission dynamics of the etiological agent of SARS in Hong Kong: Impact of public health intervention. Science. (2003), **300**:1961–1966.
[16] Marra M.A., Jones S.J., Astell C.R. et al.: The genome sequence of the SARS-associated coronavirus. Science (2003) **300**:1399–1404.
[17] MMWR 52:405 (2003) 103 patients infected by 5 sources.
[18] Fouchier R.A., Kuiken T., Schutten M. et al: Aetiology: Koch's postulates fulfilled for SARS virus. Nature (2003) **423**:240.
[19] Nicholls J.M., Poon L.L.M., Lee K.C. et al.: Lung pathology of severe acute respiratory syndrome. The Lancet (2003) **361**:1773–1778.

[20] Wenzel R.P., Edmond M.B.: Managing SARS amidst uncertainty. New England J. Med. (2003), **348**:1947–1948.
[21] Niedrig M., Lim W., Doerr H.W., Peiris M., Zambon M., Leitmeyer K., Mackenzie J., Stöhr K., Drosten C.: Results of the 1st External Quality assurance for SARS new Coronavirus diagnostic PCR and serology. International Conference on SARS, (2004) May 8–11, Lübeck.
[22] Yamamoto N., Yang R., Yoshinaka Y., Amari S., Nakano T., Cinatl J., Rabenau H., Doerr H.W., Hunsmann G., Otaka A., Tamamura H., Fujii N., Yamamoto N.: HIV protease inhibitor nelfinavir inhibits replication of SARS-associated coronavirus. Biochem. Biophys. Res. Commun. (2004), **318**:719–725.
[23] Hamming I., Timens W., Bulthuis M., Lely A., Navis G., Van Goor H.: Tissue distribution of ACE2 protein, the functional receptor for SARS coronavirus. A first step in understanding SARS pathogenesis. J. Pathol (2004), **203**:631–637.
[24] Lai M.M.C.: Recombination in large RNA viruses: Coronaviruses. Seminars in Virology (1996), 7: 381–388.
[25] Wang M.-D., Jolly A.M.: Changing virulence of the SARS virus: the epidemiological evidence. Bull. WHO. (2004), 82:547–548.

Emerging Biological Threat
G. Berencsi et al. (Eds.)
IOS Press, 2005

Antiviral Drugs for Treatment of Herpes B Virus Infections

Maria M. MEDVECZKY[a], George E. WRIGHT[b], Richard E. EBERLE[c] and Peter G. MEDVECZKY[a]
[a]*Dept. Medical Microbiology and Immunology, University of South Florida, Tampa, FL USA*
[b]*GLSynthesis, Worcester, MA USA*
[c]*Dept. Veterinary Pathobiology, College of Veterinary Medicine, Oklahoma State University*

Herpes B virus (HB, Cercopithecine Herpes Virus 1) is a primate herpesvirus that has been reported to cause lethal infections in humans. To date, about two dozen well-documented cases of human infections with HB have been reported, with mortality of 75%. HB is indigenous to macaque monkeys, and macaques used in biomedical research including both rhesus (*M. mulatta*) and cynomologus or long-tailed macaques (*M. fascicularis*) are commonly infected with this agent without showing any visible signs of disease. Following primary infection, macaques may develop oral or genital herpetic lesions much like those caused by herpes simplex virus (HSV) in humans, and the virus subsequently establishes a latent infection in sensory ganglia. Latent HB virus may reactivate in response to stress and be shed in various body fluids. HB transmission to humans usually occurs from direct contact with body fluids such as saliva from infected monkeys. Primary human disease is characterized by vesicular eruptions on the skin or mucous membranes that are indistinguishable from herpetic lesions caused by HSV1 or HSV2. Unlike typical HSV infections that rarely lead to central nervous system involvement, HB causes a rapidly ascending myelitis and encephalitis that almost always leads to death.

There are important issues of particular significance regarding Herpes B infection and bioterrorism. First, initial clinical symptoms of HB infection are similar to the ones caused by HSV1. Second, HB is about 10 times less sensitive to the standard antiviral drug acyclovir than herpes simplex virus. Therefore, victims of a Herpes B bio-terrorist attack would be likely misdiagnosed as ordinary herpes simplex cases and would be treated with standard oral doses of acyclovir. This treatment is probably not sufficient for control of infection and could lead to encephalitis and death. Furthermore, since human-to-human transmission of HB has been documented, accurate and rapid diagnosis is essential not only to initiate an effective treatment regimen but to allow isolation of infected individuals to prevent additional spread of the virus.

Therefore, two areas, diagnosis and treatment of HB infection, need to be improved. The central hypothesis of our work is that small molecules can selectively and potently inhibit HB enzymes required for the synthesis of the viral genome and can be developed as effective therapeutics for virus disease. We are currently developing new antiviral targets for HB infections involving cloning of selected DNA replication-related genes, expression and isolation of the protein products, and characterization of the substrate and inhibitor properties of the enzymes. Current antiviral drugs for the treatment of HSV infection target primarily two enzymes – the viral DNA polymerase (POL) and thymidine kinase (TK) – both of which share homology with the Herpes B enzymes. In addition, the helicase-primase (HP) complex of HSV has recently been validated as an antiviral target. Because of the nature of the pathology of Herpes B infection in humans, we are focusing on TK, because inhibitors of the HSV types 1 and 2 TKs protect mice from lethal encephalitis. Ongoing research in our laboratories include cloning and overexpression the thymidine kinase (TK) of Herpes B in eukaryotic expression vectors, testing an existing library of N^2-phenylguanine derivatives for inhibition of HB TK, and testing antiherpes drugs for activity in vitro against HB in cultured cells, in order to compare their activity against HSV, and to screen inhibitors of HB TK for antiviral activity in Vero cell cultures.

Emerging Biological Threat
G. Berencsi et al. (Eds.)
IOS Press, 2005

Tick-Borne Bacterial Diseases in Hungary

Andras LAKOS, MD, PhD
Centre for Tick-borne Diseases, Budapest, Hungary
E-mail: ALAKOS@axelero.hu

Ticks are obligate hematophagous parasites attacking every class of vertebrate throughout the world. Currently, there is an increasing awareness of tick-borne diseases. Since the identification of *Borrelia burgdorferi*, eleven tick-borne bacterial pathogens have been described in humans. This review show that Central Europe is a heavily infested area where a new tick-borne rickettsiosis was described.

1. TIBOLA (Tick-Borne Lymph-Adenopathy)

It has been demonstrated that numerous rickettsiae are maintained in and distributed by ticks. Ticks may act as reservoirs and vectors as well. Until recently only one rickettsiosis was thought to be prevalent in Europe, Mediterranean spotted fever ("boutonneuse" fever) due to *Rickettsia conorii.* During the last few years, at least four emerging tick-borne rickettsioses were described. The present data suggest that the most prevalent and distributed infection is TIBOLA, caused by *Rickettsia slovaca* and transmitted by *Dermacentor marginatus* and *D. reticulatus* ticks. (*Lakos A.: Tick-borne lymphadenopathy – a new rickettsial disease? Lancet 1997, 350, 1006.*) The tick bite in TIBOLA is almost exclusively found in the scalp region. Children are more frequently infected. The main symptoms are: eschar at the site of a tick bite, lymphadenopathy in the head and neck region, slight general symptoms, malaise, fatigue, muscle pain, rigidity of the neck and low-grade fever. Weeks after healing of the eschar, a 1–5 cm alopecia remained at the site of the tick bite. The infection is benign, only 5 cases of neurological involvement has been described throughout Europe. Most interestingly, the infection can be chronic; we have patients with 18-month-five year's disease duration. Treatment with doxycycline is able to shorten the disease, but half of the patients show spontaneous recovery. The disease is widely distributed in Southern and Central Europe from Spain to Bulgaria. Most of the cases were reported from Hungary and France. (*Raoult D., Lakos A., Fenollar F., Beytout J., Brouqui P., Fournier P.E.: Spotless Rickettsiosis caused by Rickettsia slovaca and associated with Dermacentor ticks.*) We summarised the epidemiological and clinical symptoms of 86 cases last year (*Lakos A.: Tick-borne lymphadenopathy – TIBOLA. Wien Klin Wochenshr 2002, 114, 648–54*).

2. Ehrlichioses

Human granulocytic ehrlichiosis is an infectious disease caused by the HGE agent, now taxonomically classified as *Anaplasma phagocytophilum.* The first European case in human was reported from Slovenia, in 1995. In contrast with the enormous effort to find new cases, the whole PCR-proven European HGE population consists of less than 10 patients.

Because of the difficulties in serological confirmation, the only reliable diagnostic tool is PCR.

In Hungary, we have seen several suspicious cases of HGE, but the serological results were contradictory as they were repeated in different European and US labs. PCR has not been done.

Human monocytic ehrlichiosis in the lack of the specific tick vector (*Amblyoma americanum*) does not occur in Europe. Those papers, which were published in the first enthusiastic years describing HME in Europe, were most probably based on laboratory mistakes.

3. Lyme Disease

Lyme disease is a well-known multisystem disease caused by *Borrelia burgdorferi* sensu lato. Hungary is a heavily infected area. Until 1989, the clinical and laboratory diagnosis was centralised. We have summarised the clinical and epidemiological data of these first years (*Lakos A.: Lyme borreliosis in Hungary – epidemiological analysis of 1175 cases. Report of WHO workshop on Lyme borreliosis diagnosis and surveillance. Warsaw, June 1995. WHO/CDS/VPH/95. 141–1. pp 10–11; 84–102*). Close to the 30th anniversary of describing Lyme disease, we still have to face several diagnostic and therapeutic problems.

4. Tularaemia

Tularaemia is a zoonotic disease caused by *Francisella tularensis.* A less virulent biovar (*palaeartica* or biovar B) occurs in Europe. Biovar A (or *nearctica*) rarely found in Europe. The pathogen is able to infect hundreds of animals. Ticks transmit a significant portion of sporadic cases. The vector is usually Ixodid ticks, less frequently Dermacentor sp. In Hungary, Ferencz A. collected 28 cases of tick-borne tularaemia in a period of 2 years (*Ferencz A.: Tick-borne tularaemia and its treatment. Infekt Klin. Mikr. 1997, 3–4, 116–117*).

Emerging Biological Threat
G. Berencsi et al. (Eds.)
IOS Press, 2005

Emerging or Re-Emerging Virus Diseases: Genetic Variation, Immune Failure or Human Mistake

J. RAJČÁNI and J. PASTOREK
Institute of Virology, Slovak Academy of Sciences, 84505 Bratislava, Slovak Republic

Dangerous viruses are highly contagious and pathogenic, i.e. they reach the target organ quickly causing severe morphologic changes. The tools of virus pathogenicity involve the capsid or envelope proteins that mediate adsorption and penetration to susceptible cells. In addition, virus-coded non-structural proteins trigger the shut-off of the host cell proteosynthesis and/or inhibit the host cell mRNA formation. In many cases, the virus-coded enzymes determine the rate of virus DNA or RNA synthesis, the destruction of host cell organelles and the assembly of new virus particles. The mechanisms selecting strains of higher or lower virulence among the virion progeny, follow the basic principles of genetics and apply equally to agents and their hosts. The outcome depends on the portal of entry into the human body, the route of transmission, the virulence of the agent and of its resistance to environmental conditions. Immobilization of the host by an average air-borne respiratory virus would create disadvantage for the spread of the given agent. Unless extremely contagious, it needs time to replicate and disseminates before killing the host. On the other hand, arboviruses transmitted by vectors may immobilize the host without influencing the frequency of transmission. The emerging viruses have not occurred before (HIV/AIDS, coronavirus/SARS) or they escaped attention, when affecting small and remote communities (Ebola and Lassa viruses). Some viruses, though certainly occurring before, had not been recognized to cause disease (hantavirus pulmonary syndrome, hepatitis C virus). These agents do not emerge as a result of mutations or recombinations, but rather represent a new germ, which survived in an animal reservoir. Re-emerging infections are diseases that once were a major public health problem, later on declined, but recently have become important again (reappearance of new influenza strains, dengue fever, rabies, West Nile virus and others). The control of emerging infectious agents by inhibiting the spread of particularly virulent variants of pathogens needs special control programs and surveillance systems (i.e. ProMED, Eurosurveillance). In addition, effective measures for fighting poverty and malnutrition, special vaccination and education programs, available medical care and treatment, and last but not least, alternative molecular methods for DNA or RNA identification (16S rRNA sequencing, broad range PCR and RT-PCR, representational difference analysis, novel toxin bioassays, comprehensive host gene expression profile etc.) are desirable. Increased urbanization and population density, social and political factors, local armed conflicts, increased migration and travel, natural and environmental changes and unexpected climate disasters make outbreaks of new infections possible and unpredictable.

Basic Definitions

Pathogen is an in infectious agent causing disease in virtually each susceptible host. Opportunistic pathogens cause disease in selected groups of subjects (when the immune system is failing or suppressed, in elderly, in cancer patients, in AIDS, after severe surgery or injuries etc. Contagious infectious agents are easily transmitted or they are more resistant under environmental conditions. Virulent agents are pathogens which multiply to higher titers at the portal of entry and/or quickly invade areas of the body, where they are not typically present [1]. Virulent viruses spread to target organs, the involvement of which causes disease.

The Tools of Virulence (Virulence Factors) in Bacteria

Bacteria are prokaryotic cells, which reproduce in tissues by binary fission. Their DNA has at least 1000 genes, more complicated bacteria have about 5000 genes. Several complex outer membrane structures composed of proteins and lipopolysacharides as well as capsular polysaccharides serve as tools of virulence and/or protective antigens. The ability of bacterial cells to resist phagocytosis comes from their mucous or slime layer. The protein A of staphylococci binds the Fc domain of immunoglobulins in a non-specific manner in order to interfere with opsonization. Resistance to intracellular digestion (killing) after phagocytosis is mediated by the complexity of bacterial wall (*Mycobacterium tuberculsosis*) or by the secretion of enzymes (Listeria). Outer membrane bound endotoxins (lipopolysaccharides) of gramnegative bacteria and several outer membrane proteins (such as protein M of Streptococci) exert profound biological effects on the host immune system, interact with endothelium cell receptors and disrupts the chemokine balance. Many bacteria produce toxic enzymes released into the close vicinity which are either cytocidal of inhibit chemotaxis. Production of exotoxins (proteins toxic for neural cells, leukocytes, surface epithelium cells, capillary endothelium cells) is a very potent tool of bacterial virulence. The presence of adhesive surface structures (pilins, fimbriae, outer membrane proteins) may be important for colonization of mucous membrane surfaces. An opposite feature, namely the mobility of bacterial cells due to flagellar movement, is helpful for getting access to the sites of action [2,3]. Extensive medical treatment of bacterial diseases by chemotherapeutics and antibiotics contributed to the selection resistant bacterial populations, a property frequently acquired by plasmids or transposons (i.e. by movable DNA elements). The latter property may be understood as an additional virulence factor, which developed during the last decades.

The Tools of Virulence in Viruses

The viruses are complex infectious entities equipped with own genes (ranging from less than ten to over 100) which multiply in susceptible host cells reproducing the same particles hundred or thousand fold. It comes from this definition that viruses do not have any cell-organization, but they are able to modify the events in infected cells forcing them to reproduce (100, 1000 ore even much more times) the virus particle which penetrated the cell. The glycoproteins of enveloped viruses are responsible for adsorption and penetration and are the main virulence factors. Both interact with cell receptors, which exert different functions essentially not related to adsorption. The capsid polypeptides of naked (non-enveloped) viruses posses alternative properties [4]. The pathogenicity of virus particles depends, in addition, of the efficiency of virus coded enzymes involved in replication of viral DNA or RNA (DNA polymerase, RNA polymerase, reverse transcriptase) as well as of non-structural proteins metabolizing nucleotides (thymidine kinase) or of non-structural regulatory (transactivator) proteins acting at the beginning of virus replication cycle. Mutations in the variable regions of virion proteins, which interact with cell receptors, cause antigenic changes (antigenic drift) modifying the affinity of antibodies to interfere with the surface proteins. The appearance or re-appearance of new surface glycoprotein molecules (of a different antigenic type) occur due to recombination events in animal hosts (antigenic shift).

Epidemiology Can Assess the Spread of Infectious Agents Within the Population of Man and/or Domesticated Animals

The reservoir of the infectious agent (virus, bacterium) is the site where it survives. Thus, domesticated and/or wildlife animals often serve as reservoirs of human disease. The processes which give rise to emerging infectious diseases in wildlife can be categorized as follows: 1) Ecosystem alterations of anthropogenic or natural origin; 2) Movement of pathogens or vectors, via human or natural agency; 3) Changes in microbes and/or in the recognition of emerging pathogens due to advanced diagnostic techniques. For example, hantavirus infections in rodents have emerged due to human-induced landscape alterations and/or climatic changes influencing population dynamics of hantavirus reservoir hosts, with disease consequences for humans. Rabies is an ancient disease, but its geographic expansion has occurred by both natural and anthropogenic movements of wild animals [5]. In general, viruses may be transmitted by direct or indirect contact. Direct contact includes touching or biting, inhalation of droplets, ingestion of infected food, sexual contact, contact with infectious blood, injection by contaminated syringes. Indirect contact occurs, when the agent is exposed to environmental conditions for a certain period before infecting the host. Contact by touching or by oral route occurs among children by contaminated toys, among adults by contaminated tools (including medical tools), by ingesting contaminated food or beverages, by contaminated sewage water etc. Several diseases are transmitted by infected arthropod vectors (such as mosquitoes and ticks) some other sand flies or fleas [6–8].

Route of Infection May Determine the Pathogenesis

The route of transmission may be critical for the understanding of pathogenesis. At the onset of virus replication at the portal of entry, the inoculation dose may be essential. The immediately available non-specific defense is most important within the first 12–48 hours p.i., especially during primoinfection (local IgA antibody, in-apparent skin injuries, impaired beating of kinocilia in the upper or lower respiratory tract, impaired phagocytosis in diabetes or other exhausting diseases, abnormal pH of gastric juices in some stomach diseases, the presence of lysozyme in the saliva and tears, the presence of the normal bacterial flora in the large intestine, production of interferon at early stages of infection etc.). Blood contains several proteins helping non-specific defense such as the complement system, the C-reactive protein etc. The complement system participates in the specific as well as in non-specific responses. In the latter, the alternative complement pathway is activated. Small amounts of C3b are constantly formed in the plasma to interact with bacterial and viral surfaces. Since these lack the serum regulatory factors such as *H* or *I*, causing C3 hydrolysis, the accumulation of C3b on bacterial surface may trigger the formation of C5 convertase complex [9].

The State of Immunity

When a viral (or bacterial antigen) triggers the specific immune response by repeated contact, the corresponding peptides are recognized by memory B and T lymphocytes, which begin to proliferate quickly. In such case, the specific response prevails from the very beginning. Thus, much more antibodies (with a well shaped antigen binding site) and greater amounts of cytotoxic T cells (equipped with the antigen-specific TCR) are being made. The prompt development of both arms of immune response (by activation of memory lymphocytes) explains the effectiveness of vaccination. An ultimate effectiveness of vaccination was worldwide achieved with smallpox. WHO has set forward programs for eradication of polio- and measles viruses. The wide use of HBV vaccine seems desirable as well, but has not been introduced in each country yet. Great achievements were made in preparation of an effective influenza vaccine; however, this virus is an example of rapidly changing and re-emerging pathogen, which may be difficult to eradicate by using a single measure only, i.e. by vaccination alone, since antigenic shifts may spontaneously occur. A major challenge is the control for influenza virus in animal reservoirs [10], and the analysis of re-emerged isolates. Finally, improvement of global vaccine manufacturing capacity should include the introduction of vaccines including the newly emerged H5N1 and H7N1 virus reassortants [11].

Antigen Specificity Shapes Immune Response

The variable regions of Ig molecules and of the T-cell receptors had been selected for antigens "circulating" in human population and surrounding ecosystems by the usually available infectious agents (bacteria, viruses). The emerging agent shows an unusual (new) antigenic specificity; in such case, the formation of a well fitting antibody or TCR may be difficult and time consuming task for the host immune system. Antibodies are still important but less essential in virus infections, due to the leading role of cytotoxic T cells for elimination of the virus-infected host cells [12]. Picornaviruses represent an exception from this rule. In contrast, antibodies seem up most essential for the clearance of bacteria and neutralization of their toxins. The viral antigen-antibody complex may induce anti-idiotypic antibody production, if the new complex interacts with the receptors on bystander B cells causing expansion of a new B cell clone. The anti-idiotypic antibody may cause autoimmune disease since it mimics the antigen itself.

Viruses Have Developed Mechanisms to Evade Immune Response

Lymfocytes T/CD8 recognize peptides at *de novo* proteosynthesis. The viral polypeptides in the cytoplasm bind to ubiquitin (Ub) to become digested within proteasomes. Alternatively, a special transporter protein (TAP) transfers the viral peptide to the ER, where it binds to the HLA I complex in order to become the target for cytotoxic T/CD8 lymphocytes. When macrophages or dendritic cells present the antigen from digested virions, the immunogenic peptides accumulate in the endolysosomes. These merge with vesicles derived from the endoplasmic reticulum, The virion peptides associated with the HLA II molecules, and then get transported to cell surface in order to be recognized by CD4 T lymphocytes. Viruses encode non-structural proteins, which inhibit the activity of TAP or impair HLA glycoprotein formation. This decreases the T cell mediated antigen recognition and/or the T cell cytotoxic response [13].

The interaction between IFN and corresponding IFN receptor (at the surface of neighbor non-infected cells) induces the expression of PKR (protein kinase RNA activated). This protein phosphorylates the initiation cofactor (eIF2 protein), which becomes the part of translation initiation complex. Since IFN activates PKR and reduces translation via the phosphorylated eIF2, the IFN mediated antiviral state is caused (in addition to other

Table 1. Examples how viruses evade (overcome) the action of the host immune system*.

Evasion technique	Evasion mechanism
Avoidance: the viral genetic information (provirus, latent DNA) hides from the host immune response (latency is being established).	Down regulation of the expression of structural virus genes. No proteins expressed during latency. Limited expression of non-structural proteins with insignificant antigen recognition (nuclear localization, lack of immunodominant epitopes. Mutations and reassortment yield new structural (envelope) antigens which can not be immediately recognized by primed immune cells and antibodies.
Inhibition: the virus inhibits the induction phase or the effector phases of immune response.	Infection of the thymus induces tolerance due to deletion of T cells able to recognize the given antigen(s). Infection of T or B lymphocytes and/or macrophages impairs their function (destruction of antigen presenting cells). Interference with antigen presentation. Interference with target cell destruction. Down regulation of MHC (HLA) glycoproteins.
Evasion of cytokines and antibodies	Viral proteins interfere with cytokine (and/or interferon) actions. Virus coded chemokine receptors bind chemokines (cytokines) eliminating the natural target response. Antibodies bound to virions enhances macrophage infection (dengue fever).

* From "Pathogenesis of Viral Diseases: Viral Strategies and Host Defense Mechanisms, pp. 145–160, in (Murphy, F.A., Gibbs, P.E., Horzinek, M.C., Studdert, M.J. eds.) *Veterinary Virology*, Academic Press, San Diego, 1999."

mechanisms) by active PKR formation stopping translation. HIV and many gamma herpesviruses interfere with PKR activity by encoding a regulator protein (vIRP = inteferon regulating protein) which binds PKR reversing the antiviral state [14–16].

Virus Transmission and the Host Genetic Background

Severe illness and immobilization of the patient may reduce the potential for transmission. With the exception of arboviruses, this scenario seems true for many air-born and alimentary infections. One may anticipate that convalescent virus carrier patients are frequently those, who move into contact with healthy persons. When a separated sewage system assures clean water supply rendering the water-borne fecal contamination less probable, less pathogenic enterovirus strains would become selected. In turn, picornaviruses, which are durable in the environment, may evolve a higher degree of virulence until the movement of an individual brings them into contact with the next host. Sexually transmitted pathogens usually develop chronic infections to ensure successful spread, because the period for changing the sexual partner may be relatively long (many months or years). By increasing the frequency of sexual partners, the virulence of the circulating agent may increase by selection mechanisms. For example, the virulent HIV-1 serotype E is more frequent in Thailand anticipating a rapid sexual encounter. On the other hand, HIV-2 is frequent in African populations, which reveal a lower frequency of changing the sexual partners. If tissue tropism of a virus involves lymphocytes and/or macrophages, the agent usually has the poten-

Table 2. Association between virus transmission and virulence.

Transmission	Pathogenicity (virulence)
Arthropod borne	Lethality relatively higher among vector-borne pathogens than among directly transmitted ones.
Water borne	Virulence of Salmonella, Vibrio, Shigella is higher at water borne transmission (due to fecal contamination).
Attendant borne	Virulence of E.coli, Yersinia, Clostridia correlates with the durability of environmental cycling.
Air borne	Virulence of respiratory tract pathogens correlates with durability within the environmental cycling.

tial to establish persistence (latency) from the very beginning. When later productive virus replication occurs, this causes a collapse of immune system (AIDS).

Mutation Frequency and Selective Pressure

High mutation rates have been reported for influenza genes, for spumavirus genes (*Retroviridae*) but also in the case vesiculovirus genes (*Rhabdoviridae*). The HIV protease is pepsin-like, and is similar to that of visna virus. Also retroviruses are changing rapidly. Despite of genetic drifts, many factors act to minimize the emergence of mutation-based virus epidemics.

If an entirely new reassortant virus appearing in the animal host as result of recombination events and/or due to unexpected selective pressures by accident infects humans, a influenza pandemic may develop. Tissue tropism determinants are coming from mutations in the genes encoding viral attachment proteins but also from mutations in their transcription factors. Host immunity contributes significantly, since the amounts of virus reaching the portal of entry may not be high; the virus needs at least 2–3 replication cycles before initiating the pathogenetic process. It may be expected, that dangerous viruses will emerge occasionally, however, their prediction seems difficult. Even when successfully predicted (for example the jump of prion disease such as spongiform encephalopathy to man causing infectious Jacob-Creutzfeldt disease), there is difficult to make a statement convincing the authorities.

Human Immunodeficiency Virus

HIV-1 and HIV-2 (human immunodeficiency viruses 1 and 2) are the classical examples of newly emerged agents causing devastating disease in humans. At the onset of each replication cycle, the DNA copy of vRNA (the provirus) becomes integrated into the genome of carrier T lymphocytes. HIV attacks T/CD4 cells, which are of key importance for the host immune response [18]. The variable regions of the heavily glycosylated surface HIV-1 glycoprotein (gp120) undergo mutations rapidly and show relatively low immunogenicity [19]. The HIV genome encodes several regulatory proteins promoting infectivity and activeting transcription (Tat, Rev and Vif), facilitating virion release (Vpu), increasing virus yield (Vpr) and down-regulating the MHC class I molecule synthesis (Nef). Thus, HIV reveals several alternative strategies in order to escape immune response: 1. Provirus latency; 2. MHC I inhibition; 3. Surface glycoprotein variability and 4. Chemokine receptor utilization [20]. The HIV diversity is a property of productive virus replication especially during vRNA reverse transcription and at copying the new vRNA, so that the progeny of a single

Table 3. A survey of recently emerged viruses (last 30 years).

Virus	Year	Disease entity, note
Rotavirus (*Reoviridae*)	1973	Infantile diarrhea
Parvovirus B19	1975	Chronic hemolytic anemia, aplastic crisis
Ebola virus (*Filoviridae*)	1977	Hemorrhagic fever (Ebola fever)
Hantaan virus (Hantavirus)	1978	Hemorrhagic fever with renal syndrome
HTLV 1 (Deltaretrovirus)	1980	T cell lymphoma/leukemia
HTLV-2 (Deltaretrovirus)	1982	Hairy cell leukemia
HIV (Lentivirus)	1983	Acquired immunodeficiency syndrome
HHV6 (*Beta herpesvirinae*)	1988	1. Rubeola subitum
Hepatitis C (*Flaviviridae*)	1988	NonA-nonB hepatitis (blood transmitted)
Hepatitis E (Calicivirus?)	1989	NonA-nonB hepatitis (alimentary transmitted)
Guanarito virus (*Arenaviridae)**	1991	Venezuelan hemorrhagic fever
Human metapneumovirus (HMPV)	1991	New pneumovirus (van den Hoogen et al., 2001)
Sin Nombre virus (Phlebovirus)*	1993	Acute respiratory distress syndroma
Sabia virus (*Arenaviridae*)*	1994	Brazilian hemorrhagic fever
Alkhurma virus (Flavivirus)*	1995	Hemorrhagic fever (in Saudi Arabia)
HHV8 (KS Herpes Virus)	1995	Kaposi's sarcoma (KS)
West Nile virus (Flavivirus)*	1996	Meningitis (Nedry and Mahon, 2003)*
Nipah virus (Paramyxovirus)	1998	Meniongoencephalitis (Lee et al., 1999)
A/Hong Kong/156/97(H5N1/97)	1998	Fatal "bird flu" influenza**
SARS Coronavirus (Tor2)	2003	Severe acute respiratory syndrome (SARS)

* Additional Arena-, Toga-, Bunya- and Flaviviruses have been isolated during the last decades; the number of arboviruses increased from 360 as listed in 1970 to about 520 in 2002. Out of the 100 arbovirus pathogens recognized so far, at least 20 fulfill the criteria of re-emerging viruses. Their number may not be final.
**(compare with Table 3).

Table 4. Examples of re-emerging bacterial and viral diseases.

Disease	Agent	Contributing factor(s)
Diphtheria	*Corynebacterium diphtheriae*	Interruption of immunization programs due to political reasons.
Pertussis	*Bordetella pertussis*	Refusal of vaccination, decreased vaccine efficacy, waning immunity.
Tuberculosis	*Mycobacterium tuberculosis*	Multiresistant pathogens, immunocompromized hosts (malnutrition, HIV, extreme poverty).
Yellow fever, dengue fever	Flaviviruses	Insecticide resistance, limitations for organophosphorus compounds, urbanization problems.
Pandemic influenza	Orthomyxovirus**	Periodical emergence of new serotypes, pig/duck agriculture, animal reservoirs.
Rabies	Rhabdovirus	Expensive vaccination; animal reservoirs; difficulties with eradication from wildlife.
Measles	Morbillivirus	Failure of vaccination programs, usually a lack of the second vaccine dose.
Meningitis	*Neisseria meningitidis*, Entero-(picorna)viruses, Flaviviruses, Orbiviruses, Togaviruses*	Uncertain reasons render the re-emerged virus isolates more pathogenic than were their counterparts (New mutations and/or reversions? Impaired immune response? Tropism for hemopoetic cells?).
Coxsackievirus A24 variant	Picornavirus	acute hemorrhagic conjunctivitis (VP1 gene mutations [17].

* St. Louis encephalitis virus, Eastern and Western equine viruses, Powassan virus, West Nile virus, Colorado tick fever virus (Romero and Newland, 2003).
** Goose/Guangdong/1/96 (H5N1 Gs/Gd); A/Hong Kong/156/97 (H5N1/97).

clone may reveal some heterogeneity [21]. In association with this, the viral reverse transcriptase undergoes genomic variations (10^0 out of 10^5 nucleotides) allowing the selection of mutants resistant to transcriptase inhibitors which are used as antiviral drugs [22]. The above mentioned reasons seem to explain the problems with developing a candidate vaccine capable to induce broadly reactive neutralizing antibodies and to achieve a proper cytotoxic lymphocyte response. This goal might be achieved by using the conservative domain of the surface gp120 as recombinant antigen [23].

The Hantavirus Pulmonary Syndrome (HPS)

Over 20 members of the Hantavirus genus have been isolated till now [24]. With these, two major clinical syndromes are associated, namely the hemorrhagic fever with renal syndrome (HFRS, endemic in Balkan peninsula, in Far East Russia, in Korea and China) and the cardiopulmonary syndrome (HPS, endemic in huge parts of America including southwestern USA). The HPS was first recognized in 1993 associated with deer mouse (*Peromyscus*) [25]. Hantaviruses (*Bunyaviridae* family) have 3 vRNA segments of minus polarity, termed L, M a S; from these, the M encodes two surface glycoproteins, G1 and G2. During replication, the vRNA segments are copied by the viral *RNA polymerase* to prepare full length cRNA segments, which serve as templates for vRNA synthesis. The mRNAs are made by adding the capped 5'-terminus primers (coming from cellular mRNAs) to the subgenomic plus RNA transcripts by a mechnanism called "cap snatching". The HPS virus replicates in the lung capillary endothelium cells [26]. The divergence of hantavirus surface glycoproteins provides some insight into the interaction of hantaviruses with endothelium cells. The G1 and G2 glycoprotein variations are likely to contribute to interactions that determine pathogenic responses to individual hantaviruses. [27]. Pathogenic hantaviruses appear to suppress early cellular IFN responses that are activated by nonpathogenic hantaviruses such as Prospect Hill virus. During HPS, there is little immune cell recruitment to the site of infection, whereas immune responses and immune complexes have been implicated in the HFRS disease process. At late times p.i., the vascular endothelium commonly induces about 13 cellular genes triggered by both HPS as well as HFRS strains. In addition, HFRS strains (but no t the HPS strains) uniquely induce a variety of chemokines and cell adhesion molecules (i.e., IL-8, IL-6, GRO-β, ICAM), as well as two complement cascade-associated factors that may contribute to immune components of HFRS disease [28]. The cDNA prepared by reverse transcription from the S segment, amplified and biotin labeled by PCR, was used with success for demonstration of vRNA in disease reservoirs such as tats (*Oryzomys, Sigmodon*) and mice (*Calomys, Peromyscus*) [29].

Lassa Virus is an Arenavirus

Lassa fever (endemic in Africa) is a severe systemic illness, associated with massive hemorrhages and a shock syndrome. Involvement of macrophages is extensive, platelets show aggregation defects, no intravascular coagulation but hemoconcentration may develop [30]. Virus isolation is recommended from blood samples. Meningeal symptoms are infrequent. Convalescent patients secrete virus in semen and urine. Increasing international travel and the possibility of use of the Lassa virus as a biological weapon escalate the potential for harm beyond the local level [31]. Several other hemorrhagic fevers attributed to emerging viruses (Guanarito in Venezula, Sabia in Brazil) were classified as Arenaviruses. All these infections are typical zoonoses (Machupo virus HF in Bolivia, Junin virus HF in Argentina, Aroyo in North America, LCM virus causes fever and meningitis worldwide). However, no hemorrhagic syndrome develops in LCM infection.

Arenaviruses have an ambisense RNA organized in two segments, L and S. The L segment encodes the large L protein (RNA polymerase), while the S segment encodes the N protein and the surface GP. Ribosomes seen within the virion by EM, yielded the name for this family (arena = sand). Each segment uses an ambisense coding organization to direct the synthesis of two gene products in opposite orientation, and each is separated by an intergenic region [32] so that the transcription strategy differs by different genes. For example, the GP precursor polypeptide, from which the two surface glycoproteins (GP-1 and GP-2) are derived, is transcribed from the anti-genomic RNA of the S segment. The corresponding ORF of the genomic RNA has a plus polarity. In contrast, the N mRNA is transcribed from the minus sense portion of the genomic RNA from the same segment. GP-1 and GP-2 form the spikes on the virion envelope and mediate cell entry by interaction with the host cell surface receptor [33]. The L protein (RNA-dependent RNA polymerase) becomes associated with the virions. In infected cells, a small RING-finger Z protein, encoded by the L segment, was found to represent the main driving force of virus budding [34]. The Z protein, which is a structural component of the virus, has been reported to interact with several cellular factors. Both LCMV and LFV Z proteins recruited to the plasma membrane Tsg101, which is a component of the class E vacuolar protein sorting machinery that has been implicated in budding of both HIV and Ebola viruses.

The immune tolerance can be established after congenital or perinatal infections. When infecting adults, cytotoxic T cells cause diffuse meningitis. While the T-cell mediated immunity is high, the levels of neutralizing antibodies remain low. The respiratory burst within the polymorphonuclear leukocytes is impaired (a *f-Met-Leu-Phe* chemotaxis antagonist is produced). Vaccination (recombinant GP vaccine) is availabe; rodents should be eliminated from contact with households. *Eating the meat from rodents should be discouraged* [31].

The Molecular Biology and Pathology of Filovirus Infection

Ebola virus minus sense RNA has 7 genes, separated by non-coding (IR) sequences. The editing site within the GP gene (a single A inserted) causes frameshift reading by translation resulting into two glycoproteins: the membrane anchored full length GP and the secreted sGP. Due to sGP secretion, the Ebola virus antigen can be detected in skin biopsy specimens and corneal smears. The antigen is present in the serum [35]. The virus is present in macrophages, lymph nodes, spleen, liver and blood leukocytes. Multiple foci of hemorrhage occur at mucosal surfaces, especially in the GIT, genitourinary and respiratory tracts. During development of the disseminated intravascular coagulation (DIC) syndrome, TNFα destroys endothelium cells permitting leakage of macromolecules. Body fluids, blood and secretions are infectious at direct contact. Aerosols are not extremely contagious with the exception of laboratory infections. Primates and bats were suggested as reservoirs. In clinically overt cases, the mortality is high. Since the abortive cases and inapparent seroconversions are usually not registered, the real mortality and/or morbidity rates remain unknown [36]. When Ebola outbreak occurs in central Africa, a limited person to person spread takes place not sparing the nursing personnel. More recently, using an ELISA, serum samples from Pygmy and non-Pygmy populations were tested for Ebola-Zaire virus and Marburg virus antibody. Positive sera were found in all zones investigated, and in all populations studied (Ebola virus IgG 5.3%; Marburg virus IgG 2.4%) testifying spontaneous circulation of less pathogenic variants or yet unknown low virulent but cross reacting filovirus strains [37].

The SARS Coronavirus (CoV)

The SARS-CoV has a plus sense vRNA nearly 30 kbp long. Proteins common with other coronaviruses are the S – (spike, E2), the M (transmembrane, E1), the N protein (nucleoprotein) and the RNA dependent RNA poly (Rep, replicase); the latter is non-structural and its mRNA is translated directly from the plus sense genome. An HE (hemagglutinin esterase) was also identified. The individual genes (ORFs) on the vRNA are separated by non-coding sequences equipped with an intergenomic (IG) transcription regulation signal (TRS). The candidate TRS signal sequences were scored strong and week. With the exception of *Rep* polyprotein, at least 6–7 "nested" mRNAs are transcribed from corresponding minus strand cRNAs of various lengths. All transcripts end at a common 3'-UTR (termination signal). Possibly, each mRNA starts at different IG (TRS) sequence. The 5'-end of each mRNA gets completed by a mechanism called leader jump, when attached to the leader sequence (5'-UTR) of the vRNA. The SARS-CoV has 14 ORFs, some of them encoding proteins of unknown function (X1…X5), two endogenous proteases were identified. Putative inhibitors of these proteases (one picornavirus-like, one papain-like), might be helpful anti-CoV drugs [38,39].

Pathogenesis and Epidemiology of SARS/CoV

SARS CoV replicates in the lung alveolar epithelium cells (inclusion bodies and viral paticles within cytoplasm). The alveoli contain exudates, mononuclear cells and hyaline membranes.

Massive necrosis was found in spleen and lymph nodes. This was accompanied with vasculitis and inflammation in liver, kidneys, adrenal glands, striated and heart muscles, i.e. SARS. It seems to be a systemic disease involving many organs. In mild cases the virus replicates in lungs and intestinal mucosa being excreted by respiratory tract air droplets, stool and urine. The virus can be isolated in HeLa, Vero E6 and MDCK cells. By EM, the cytoplasm contains membrane whorls and vacuole-bound virions. The existence of two SARS CoV diseases, a mild intestinal disease (A) and a severe, respiratory disease (B) was suggested. Hypothetically, the mild disease might protect from the severe one. SARS had developed in areas untouched from the "natural" (silent or mild) form of the disease (A). Whether the severe syndrome developed due to a mutant virus, should be further analyzed. It is highly probable, that the new CoV originated from the animal reservoir [40,41]. The SARS outbreak was a reminder that new infectious diseases are always dormant somewhere. In November 2002, when the first cases were registered in Guandong province, the authorities remained silent for several months contributing to the initial spread of the disease by sticking to the first hypothesis that it was a chlamydial disease. But at the end the Chinese epidemic (about 8000 reported cases, 774 deaths), the disease was brought under control by classical measures such as strict disinfections and isolation of the patients [42]. The disease spread outside China, since as of July 11, 2003, 8437 people in 32 countries have been affected, with 813 deaths reported. Fortunately, no catastrophic pandemic but rather several limited epidemics have occurred [43].

References

[1] Bomsel M. and Alfsen A: Virus entry through the epithelial barrier. *Nature Rev Mol Cell Biol* 4, 57–68, 2003.

[2] Hornef MW, Wick MJ, Rhen M, Normark S. Bacterial strategies for overcoming host innate and adaptive immune responses. *Nat Rev. Immunol.* 2002 Nov; 3(11): 1033–40.

[3] Pieters J. Evasion of host cell defense mechanisms by pathogenic bacteria. *Curr Opin Immunol.* 13(1): 37–44, 2001.
[4] Rajčáni, J.: Molecular mechanisms of virus spread and virion components as tools of virulence. *Acta Microbiol. Immunol. Hung.* 50, 407–431, 2003.
[5] Williams ES, Yuill T, Artois M, Fischer J, Haigh SA.: Emerging infectious diseases in wildlife. *Rev Sci Techn* 21(1): 139–57, 2002.
[6] Haydon DT, Cleaveland S, Taylor LH, Laurenson MK. Identifying reservoirs of infection: a conceptual and practical challenge. *Emerg Inf. Dis.* 8(12): 1468–73, 2002.
[7] Kuiken T, Fouchier R, Rimmelzwaan G, Osterhaus A. Emerging viral infections in a rapidly changing world. *Curr Opin Biotechnol.* 14(6): 641–6, 2003.
[8] Lashley FR. Factors contributing to the occurrence of emerging infectious diseases. *Biol Res Nurs.* 4(4): 258–67, 2003.
[9] Hirsch, RI: The complement system: its importance in the host response to viral infection. *Microbiol. Rev.* 46, 71–85, 1982.
[10] Webby, RJ, Webster, RG.: Are we ready for pandemic influenza? *Science* 302 (5650), 1519–1522, 2003.
[11] Guan, Y, Peiris, JSM, Lipatov, AS, Ellis, TM, Dyrting, KC, Krauss, S, Zhang, LJ, Webster, RG, and Shortridge, KF: Emergence of multiple genotypes of H5N1 avian influenza viruses. *Proc Natl Acad Sci USA.* 99 (13): 8950–8955, 2002.
[12] Whitton, JL, Oldstone, MBA: The immune response to viruses, pp. 285–320, in Knipe, D, Howley, PM (eds.) *Fundamental Virology* 4th ed., Lippincott Williams and Wilkins, Philadelphia, 2001.
[13] Bauer D, Tampe R. Herpes viral proteins blocking the transporter associated with antigen processing TAP – from genes to function and structure. *Curr Top Microbiol Immunol.* 269: 87–99, 2002.
[14] Grandvaux N, tenOever BR, Servant MJ, Hiscott J. The interferon antiviral response: from viral invasion to evasion. *Curr Opin Infect Dis.* 2002 Jun; 15(3): 259–67.
[15] Katze MG, He Y, Gale M Jr. Viruses and interferon: a fight for supremacy. *Nat Rev Immunol.* 2(9): 675–87, 2002.
[16] Leib, DA. Counteraction of interferon-induced antiviral responses by herpes simplex viruses. Curr Top Microbiol Immunol. 2002; 269: 171–85.
[17] Oh MD, Park S, Choi Y, Kim H, Lee K, Park W, Yoo Y, Kim EC, Choe K.: Acute hemorrhagic conjunctivitis caused by coxsackievirus A24 variant, South Korea, 2002. *Emerg Infect Dis.* 9, 1010–2, 2003.
[18] Douek DC. Disrupting T-cell homeostasis: how HIV-1 infection causes disease. *AIDS Rev.* 5(3): 172–7, 2003.
[19] Wyatt, R, Kwong, PD, Desjardins, E, Sweet, RW, Robinson J, Hendrickson, WA, Sodroski JG: The antigenic structure of the HIV gp120 glycoprotein. *Nature* 393, 705–711, 1998.
[20] Guatelli, JC, Siliciano, RF, Kuritzkes, DR, Richman, DD: Human immunodeficiency virus, pp. 685–729, in: Richman, DD, Whitley, RJ, Hayden, FG (eds.): *Clinical Virology,* 2nd ed., ASM Press, Washington, 2002.
[21] Shankarappa, R, Margolick, JB, Gange, SJ, Rodrigo, AG, Upchurch, D, Farzadegan, H, Gupta, P, Rinaldo, CR, Learn, GH, He, X, Huang, XL, Mullins, JI: Consistent evolutionary changes associatyed with the progression of human immunodeficiency virus type 1 infection. *J Virol.* 73, 10,489–10,502, 1999.
[22] Holland, JJ, delaTorre, JC, Steinbauer, DA: RNA virus subpopulations as quasispecies. *Curr Topics Microbiol. Immunol.* 176, 1–20, 1992.
[23] Mathieson, BJ: HIV vaccines: milestones and new insights, pp. 421–457, in (Stanberry, LR, Bernstein, DI (eds.), *Sexually transmitted diseases. Vaccines, Prevention and Control,* Ascademic Press, San Diego, 2000.
[24] Mertz, GJ: Bunyaviridae: Bunyaviruses, Phleboviruses, Nairoviruses and Hantaviruses, pp. 921–969, in in: Richman, DD, Whitley, RJ, Hayden, FG (eds): *Clinical Virology,* 2nd ed., ASM Press, Washington, 2002.
[25] Hjelle, B, Jenison, S, Torrey-Martinez, N, Yamada, T, Nolte, K, Zumwalt, R, MacInnes, K, Myers, G: A novel hantavirus associated with an outbreak of fatal respiratory disease in the southwestern United States: evolutionary relationships to known hantaviruses. *J Virol.* 68, 592–596, 1994.
[26] Zaki SR, Greer PW, Coffield LM, Goldsmith CS, Nolte KB, Foucar K, Feddersen RM, Zumwalt RE, Miller GL, Khan AS: Hantavirus pulmonary syndrome. Pathogenesis of an emerging infectious disease. *Am J Pathol.* 146(3): 552–79, 1995.
[27] Mackow ER, Gavrilovskaya IN: Cellular receptors and hantavirus pathogenesis. *Curr Top Microbiol Immunol.* 256: 91–115, 2001.
[28] Geimonen, E, Neff, Sh, Raymond, T, Kocer, SS, Gavrilovskaya, IN, Mackow ER: Pathogenic and nonpathogenic hantaviruses differentially regulate endothelial cell response. *Proc Natl Acad Sci USA.* 99 (21): 13837–13842, 2002.
[29] Lednicky, JA: Hantaviruses. a short review. *Arch Pathol Lab Med.* 127(1): 30–5, 2003.
[30] McCormick, JB, Webb, PA, Krebs, JW, Johnson, KM, Smith, ES: A prospective study of the epidemiology and ecology of Lassa fever. *J Infect Dis* 155, 437–444, 1986.

[31] Richmond JK, Baglole DJ. Lassa fever: epidemiology, clinical features, and social consequences *Brit. Med. J* 327(7426): 1271–5, 2003.

[32] Buchmaier, M.J., Bowen, M.D. & Peters, C.J. Arenaviruses, pp. 1635–1668, in Knipe, D.M. & Howley, P.M. (eds.): *Fields Virology*, 4th ed, Vol. 2, Lippincott, Williams & Wilkins, Philadelphia (2001).

[33] Cao, W., Henry, M.D., Borrow, P., Yamada, H., Elder, J.H., Ravkov, E.V., Nichol, S.T., Compans, R.W., Campbell, K.P. & Oldstone, M.B. Identification of α-Dystroglycan as a Receptor for Lymphocytic Choriomeningitis Virus and Lassa Fever Virus. *Science* **282,** 2079–208, 1998.

[34] Perez M, Craven RC, de la Torre JC.: The samall RING finger protein Z drives arenavirus budding: implications for antiviral strategies. Proc. Natl. Acad. Sci. US. 100 (22): 12978–12983, 2003.

[35] Feldmann H, Klenk HD, Sanchez A. Molecular biology and evolution of filoviruses. *Arch Virol* (Suppl.) 7: 81–100, 1993.

[36] Georges-Courbot, M, Sanchez, A., Lu, C, Baize, S, Leroy E, Lansouty-Soukate, J, Tevi-Beniossan, C, Georges, A, Trappier, S, Zaki, S, Swanepoel, R, Leman, P, Rollin, P, Peters, C, Nichol, S, Ksiazek, T: Isolation and phylogenetic characterization of Ebola virus causing different outbreaks in Gabun. *Emerg. Infect. Dis*. 3, 59–62, 1997.

[37] Gonzalez, JP Nakoune, E Slenczka, W, Vidal P, Morvan JM: Ebola and Marburg virus antibody prevalence in selected populations of the Central African Republic. Microbes Infect. 2(1): 39–44, 2000.

[38] Ruan, Yi J, Wei, ChL, Ling, AE, Vega, VB, Thoreau, H, Thoe, SYS, Chia, J-M, Ng P, Chiu PK, Lim L, Zhang, T, Chan , KP, Oon, LEL, Ng, ML, Leo, SY, Ng, LFP, Ren, ECh, Stanton, LW, Long, PM, Liu, ET: Comparative full-length genome sequence analysis of 14 SARS coronavirus isolates and common mutations associated with putative origins of infection.: *Lancet*. 361(9371): 1779–85, 2003.

[39] Kuiken T, Fouchier RA, Schutten M, Rimmelzwaan GF, van Amerongen G, van Riel D, Laman JD, de Jong T, van Doornum G, Lim W, Ling AE, Chan PK, Tam JS, Zambon MC, Gopal R, Drosten C, van der Werf S, Escriou N, Manuguerra JC, Stohr K, Peiris JS, Osterhaus AD.: Newly discovered coronavirus as the primary cause of severe acute respiratory syndrome. Lancet. 362 (9380), 263–70, 2003.

[40] Zeng FY, Chan CW, Chan MN, Chen JD, Chow KY, Hon CC, Hui KH, Li J, Li VY, Wang CY, Wang PY, Guan Y, Zheng B, Poon LL, Chan KH, Yuen KY, Peiris JS, Leung FC.: The complete genome sequence of severe acute respiratory syndrome coronavirus strain HKU-39849 (HK-39). *Exp Biol Med.* 228(7): 866–73, 2003.

[41] Tsui PT, Kwok ML, Yuen H, Lai ST: Severe acute respiratory syndrome: clinical outcome and prognostic correlates. *Emerg Infect Dis.* 9(9): 1064–9, 2003.

[42] Enserink, M: Breakthrough of the year: SARS, a pandemic prevented. *Science* 302 (5653): 2045, 2003.

[43] Whitby N, Whitby M.: SARS: a new infectious disease for a new century. Aust Fam Physician. 32(10): 779–83, 2003.

PART 2

DEVELOPMENT OF THE DIAGNOSTIC TECHNOLOGY TO COPE WITH EMERGING THREAT MORE EFFICIENTLY

Emerging Biological Threat
G. Berencsi et al. (Eds.)
IOS Press, 2005

Real-Time PCR for Detection of Parenterally Transmissible Viruses

Mario POLJAK, Boštjan J. KOCJAN, Katja SEME and Dunja Z. BABIČ
Institute of Microbiology and Immunology, Medical Faculty, Zaloška 4, 1000 Ljubljana, Slovenia

Although commercially available nucleic acid amplification assays exist for three most important parenterally transmissible viruses, e.g. human immunodeficiency virus type 1 and hepatitis B and C viruses, the relatively high price of these assays obviates their use in many parts of the world. Among recent technical improvements in the field of molecular virology, real-time polymerase chain reaction (PCR) is one of the most exciting and is expected to be an alternative and a cheaper diagnostic tool for the detection of parenterally transmissible viruses. Real-time PCR allows a simultaneous amplification and quantification of specific nucleic acid sequences. This is achieved by a combination of rapid thermal cycling and cycle-by-cycle basis detection of the reaction kinetics by means of fluorimetry.

Introduction

Among recent technical improvements in the field of molecular virology, real-time polymerase chain reaction (PCR) is one of the most exciting and is expected to be the main diagnostic molecular tool for the detection of those viruses for which no commercial diagnostic tests are available or the so-called commercially uninteresting viruses. Additionally, the real-time PCR represents an alternative and a cheaper diagnostic tool for the detection of commercially interesting viruses, including parenterally transmissible viruses.

Real-Time PCR

Real-time PCR allows a simultaneous amplification and quantification of specific nucleic acid sequences. This is achieved by a combination of rapid thermal cycling and cycle-by-cycle basis detection of the reaction kinetics or real–time detection by means of fluorimetry. Cycle-by-cycle monitoring very precisely identifies the cycles in which PCR is in the log-linear phase, with the PCR products doubling with each cycle. In this phase, the signal is easily distinguished from the background signal, providing very accurate information about the starting concentration of the target sequence. Real-time PCR is a technology in wide expansion, offering an increasing range of possibilities in instrumentation as well as real-time PCR detection chemistries [1,2].

1.1. Real-Time PCR Instruments

Real-time PCR instruments are basically fluorescence-detecting thermocyclers. At present, it is possible to choose between a variety of competing instruments [2–4].

ABI PRISM 7700 Sequence Detection System (Applied Biosystems, Foster City, CA) was the first commercially available thermocycler for real-time PCR. The system consists of a 96-well thermal cycler, precision optics, computer, and application software. Reaction tubes with transparent lids allow the light from the laser, carried on fiber optic cables, to excite the probe; the emitted light is then sent back to a CCD camera for the detection of signal. Less expensive alternative is Gene Amp 5700 Sequence Detection System that has a halogen lamp instead of a laser and allows a single-wave length detection only. Recently Applied Biosystems launched the completely automated ABI PRISM 7900HT, which has the same specification as the ABI PRISM 7700, but is designed especially for high throughput applications.

LightCycler System (Roche Diagnostics, Manheim, Germany) is beside ABI Sequence Detection Systems the most widely used thermocycler for real-time PCR. It holds up to 32 samples and uses heated air and 20 μl borosilicate glass capillaries with a high surface-to-volume ratio, allowing complete PCR to be completed in 20 to 30 minutes. The optical unit contains a light-emitting diode and three detection channels (530 nm, 640 nm, 710 nm). The new LightCycler 2.0 instrument consists of 6 detection channels and provides fluorescence detection at 530, 560, 610, 640, 670, and 710 nm. The availability of the new 100 μl capillaries offers the choice of two reaction volumes for standard or high-sensitivity applications.

Smart Cycler System (Cepheid, Sunnyvale, CA) consists of 16 I-CORE modules that represent a complete, independent, temperature-controlled fluorimeters for performing and continuous monitoring of PCR. Each module includes a powerful, four-channel optical analysis system capable of detecting and quantifying multiple fluorescent dyes and multiple target molecules in the same reaction tube. Up to 6 processing blocks can be interconnected, thus representing 6×16 independently-controlled reaction sites. An advantage of this system is its high flexibility since different PCR runs can be carried out at the same time for individual experiments.

iCycler iQ Real-Time PCR Detection System (Biorad, Hercules, CA) is a standard 96-well capacity thermal cycler coupled with data acquisition system for performing real time quantitative PCR. It covers wide range of excitation/emission wavelengths (400–700 nm) and thus facilitates a great array of fluorescent PCR strategies. Alltogether 96 samples can be tracked simultaneously and up to 4 different fluorophores can be multiplexed per sample tube.

Mx4000 (Stratagene, La Jolla, CA) represents improvement of the older Mx3000 cycler. Mx4000 is designed specifically for multiplexed PCR applications and supports all available fluorescent chemistries. This 96-well capacity thermal cycler uses single tube or tube stripes and its optical system allows to use a wide selection of fluorescent dyes. For multiplex applications up to four different dyes can be used.

Rotor Gene (Corbett Research, Mortlake, Australia) is a centrifugal thermal cycler comparable to LightCycler capable to excite and detect 4 different fluorescent dyes simultaneously. The system is supplied with 2 different rotors, (36 and 72 well formats) and is compatible with all current PCR real-time detection chemistries.

1.2. Real-Time PCR Detection Chemistries

All real-time PCR chemistries use fluorescent dyes and combine the processes of amplification and detection of PCR products. Each method differs in specificity and sensitivity and it is worth selecting the most appropriate chemistry for each specific application.

The simplest real-time PCR chemistry employs a dye with the fluorescence increasing when bound to double strand DNA (dsDNA) and two unlabeled primers without any probe. SYBR Green I dye binds to the minor groove of the DNA double helix. The fluorescence of dye increases exponentially as it binds to dsDNA. At the start of the amplification, there is little dsDNA, therefore, the fluorescence of the background is low. After primer extension is initiated, dsDNA begins to form, allowing more dye to bind in proportion to the amount of dsDNA generated. This, in turn, generates an increase in fluorescence. When dsDNA is melted for the next PCR cycle, the dye is released and fluorescence decreases. By the analysis of melting curve and quantification of the amount of fluorescence, a specific PCR product identification can be achieved. The main disadvantage of SYBR Green chemistry is non-specific binding since the dye will bind to any dsDNA sequence, including none specific products and those from primer dimers.

Since the use of dsDNA binding dyes results in a non-specific detection of amplification, alternative chemistries have been developed that utilize Föster Resonance Energy Transfer (FRET). The TaqMan real-time PCR chemistry utilizes the 5'–3' exonuclease activity of Taq polymerases to hydrolyze a hybridization probe bound to its target amplicon. TaqMan PCR uses unlabeled forward and reverse primers and a labeled internal hybridization probe. The hybridization probe is a linear oligonucleotide blocked at the 3' end to prevent extension during PCR. The 5' end of the probe is labeled with a fluorescent dye (called reporter dye) and the 3' end with a corresponding FRET acceptor molecule (called quencher). While the probe is intact, and both dyes are in close proximity, the quencher absorbs the fluorescence of the reporter dye. The probe hybridizes to a specific template sequence at a higher temperature than the primers. During amplification and primer extension, the 5' nuclease activity of the Taq polymerase degrades the probe and releases the two dyes. The two dyes are no longer in close proximity, and the reporter dye is no longer quenched. The amount of fluorescence is directly proportional to the amount of target DNA generated during the PCR process.

The third most frequently used real-time PCR chemistry, called hybridization probe chemistry, utilizes two labeled oligonucleotide probes and unlabelled forward and reverse primers. The first probe carries a donor fluorophore dye at its 3' end, whereas the second probe carries an acceptor fluorophore dye (e.g. LC Red 640) at its 5' end. The sequences of the two probes are selected in the way that they can hybridize to the amplified DNA fragment in a head-to-tail arrangement, thereby bringing the two dyes into close proximity. Fluorescein excited by the light emits green fluorescent light at a slightly longer wavelength. When the two dyes are in close proximity, the energy thus emitted excites the LC Red 640 attached to the second probe that subsequently emits red fluorescent light at an even longer wavelength. FRET is possible only if the two dyes are in close proximity (between 1–5 nucleotides). In addition to three described real-time PCR detection chemistries, real-time PCR protocols using molecular beacons, Scorpion probes and minor groove binding oligonucleotides have been recently developed [reviewed in 1, 5].

2. Real-Time PCR for the Detection of Parenterally Transmissible Viruses

In the acute phase of viral infection during viremia, theoretically most, if not all, viruses can be transmitted parenterally. However, in the present review, we will focus on five most important parenterally transmissible viruses: human immunodeficiency viruses (HIV) type 1 and type 2, human T-cell leukemia/lymphoma virus type I (HTLV-I), hepatitis B virus (HBV) and hepatitis C virus (HCV). Although excellent commercially available nucleic acid amplification assays exist for three out of five most important parenterally transmissible viruses e.g. HIV-1, HCV and HBV, the relatively high price of these tests obviates their use in many parts of the world, including some European countries which have been severely affected by HIV and hepatitis epidemics in the last 5 years. This disadvantage of commercially available nucleic acid amplification assays can be overwhelmed by using one of the numerous recently developed real-time PCR protocols. The authors selection of the most reliable and useful real-time PCR protocols for the detection of five parenterally transmissible viruses are described below in brief [2].

2.1. Human Immunodeficiency Virus Type 1 Infection

Commercially available assays for measuring the plasma HIV-1 RNA viral load are currently used to predict the progression of patients to AIDS, to monitor HIV-1-infected patients receiving potent antiretroviral therapy, and to assess the efficacy of new antiretroviral regimens to treat HIV-1 infection [3]. In-house real-time PCR assays using different chemistries can also be applied for plasma viral HIV-1 RNA [6,7] or/and proviral HIV-1 DNA quantification, thus offering an alternative tool for viral load estimation. The lack of commercially available assay that would efficiently quantify HIV-1 subtype O as well as the divergent strains of HIV-1 M group mainly enforces the development of these in-house assays [7,8]. Prolonged suppression of HIV-1 replication by antiretroviral therapy leads to dramatic reductions in HIV-1 RNA in plasma below detectable levels (20 copies/ml). Antiretroviral treatment also decreases proviral HIV-1 DNA in peripheral blood mononuclear cells, lymphoid tissues and semen, but the rate of decline is generally lower than that of plasma HIV-1 RNA. Detectable proviral HIV-1 DNA persists even after prolonged treatment and serves as reservoir for chronic infection [3,9,10]. Several highly reproducible and accurate assays for quantification of proviral HIV-1 DNA have been described with the detection limit of 10–1000 proviruses per 100.000 cell equivalents. Although the majority of the published protocols include probes/primers for B subtype, some can multiply other HIV-1 subtypes. However, it should be stressed that no single primer/probe combination is likely to amplify all clinical HIV-1 isolates because of the huge HIV strain diversity and rapid sequence evolution [3,9–18].

Since commercially available HIV-1 RNA viral load assays are approved only for monitoring of HIV-1-infected patients, in-house real-time PCR protocols are used frequently for exclusion/confirmation of HIV-1 infection in cases of indeterminate results of confirmatory serologic tests (Western-blot or immunoblot assays) and in children born from HIV-1 infected mothers.

2.2. Human Immunodeficiency Virus Type 2 Infection

HIV-2 differs from HIV-1 by its lower pathogenicity and higher level of intrasubtype strain diversity. Although HIV-2 infection is associated with plasma HIV-2 RNA viral loads significantly lower than those found in HIV-1 infection, HIV-2 DNA proviral load is relatively high, with the values similar to the HIV-1 proviral load. High level of genetic distance between the two major subtypes, subtypes A and B, makes it difficult to develop a single PCR protocol to detect their genomes. An extremely low viral load in most patients further hinders the development of a sensitive assay for HIV-2 RNA detection. However, using real-time PCR based on LightCycler HIV-2 proviral DNA of both subtypes can be now easily quantified. The recently developed assay has a detection limit of 5 copies/10^5 peripheral blood mononuclear cells and is insensitive to HIV-2 strain variability [19]. A similar LightCycler based real-time PCR method described by Damond et al. can be used also for HIV-2 RNA quantification [20]. This assay provides good reproducibility and 100% sensitivity at a viral load of 250 copies/ml plasma. Detection and quantification of HIV-2 RNA can also be done with real-time PCR assay based on TaqMan chemistry with the detection limit of $5x10^2$ HIV-2 RNA copies/ml of plasma [21].

2.3. Human T-Cell Leukemia/Lymphoma Type I Infection

The HTLV-I is another transfusion-transmissible retrovirus targeting T lymphocytes. Since no commercial PCR based assay for the detection of this virus is available, real-time PCR assay can be used for the exclusion/confirmation of infection (mainly in the cases with indeterminate serology results), distinguishing between the infections with HTLV-1 and HTLV-2 or for monitoring the patients with already established infection. Several real-time PCR quantitative assays can be used to measure the level of the integrated viral genome of HTLV-1 in host peripheral blood-mononuclear cells or whole blood samples. In almost all protocols, at least 10 copies of HTLV-1 DNA can be detected with 100% sensitivity [22–26]. Some of the authors use multiplex RT-PCRs for simultaneous detection of HTLV-1/2 [27] or all pathogenic retroviruses [11]. Although the *pol*, *gag*, *tax* and *rex* genes are the most frequently used regions for the HTLV-1 DNA detection, one should be careful when choosing regions other than *pol* [25].

2.4. Hepatitis B

Quantitation of HBV DNA in serum is a useful method for monitoring HBV replication and is essential for monitoring the efficacy of the antiviral treatments, detecting the occurrence of the drug resistant mutants and relapses after discontinuing antiviral therapy [28,29]. Low per-sample assay costs, rapidness, high sensitivity, reproducibility and more accurate measurement of levels of viral replication have render the real time PCR based assays an excellent alternative to existing, widely used commercial PCR and non-PCR based methods [29–31]. Ho et al. [29] developed LightCycler based assay that has a broad quantification range from 250 to 2.5×10^9 copies of HBV DNA per ml. Similar results were obtained using the TaqMan chemistry and ABI PRISM cycling devices [28,30–34]. Some authors use real-time PCR for studying different mutations in HBV genome, including lamivudin-resistance-associated mutations [35–37] and quantifying HBV DNA in tumorous and surroundings tissue from the patients with hepatocellular carcinoma [38], and thus expanding the usefulness of real-time PCR to other fields.

2.5. Hepatitis C

As in the case of hepatitis B, real-time PCR represents a cheaper alternative to commercially available HCV RNA qualitative as well as quantitative assays. Some of the real-time PCR assays are developed for the use on the LightCycler instrument with SYBR Green or/and FRET chemistries with the sensitivity down to 28 copies of HCV RNA per amplification reaction [39–42], while others use the TaqMan technology or molecular beacons and ABI PRISM sequence detection system. The latter has the sensitivity between 10–1,000 copies of HCV RNA per reaction [42–48]. In-house real-time PCR protocols have satisfactory sensitivity and reproducibility, broad dynamic range and show good correlation with commercial assays. Since the decision on the duration of current combination therapy is mainly based on HCV genotype, PCR determination of HCV genotype has become increasingly important. Expensive commercial methods for HCV genotype determination can be now satisfactorily replaced with the fast and reproducible real-time PCR assays as described by Schroter et al. [41] and Bullock et al. [49].

References

[1] Giulietti A, Overbergh L, Valckx D, et al. An overview of real-time quantitative PCR: applications to quantify cytokine gene expression. Methods 2001; 25: 386–401.

[2] Mackay IM, Arden KE, Nitsche A. Real-time PCR in virology. Nucleic Acids Res 2002; 30: 1292–305.

[3] Kostrikis LG, Touloumi G, Karanicolas R, et al. Multicenter Hemophilia Cohort Study Group. Quantitation of human immunodeficiency virus type 1 DNA forms with the second template switch in peripheral blood cells predicts disease progression independently of plasma RNA load. J Virol 2002; 76: 10099–108.

[4] Ong YL, Irvine A. Quantitative real-time PCR: a critique of method and practical considerations. Hematology 2002; 7: 59–67.

[5] Klein D. Quantification using real-time PCR technology: applications and limitations. Trends Mol Med 2002; 8: 257–60.

[6] Lewin SR, Vesanen M, Kostrikis L, et al. Use of real-time PCR and molecular beacons to detect virus replication in human immunodeficiency virus type 1-infected individuals on prolonged effective antiretroviral therapy. J Virol 1999; 73: 6099–103.

[7] Gueudin M, Plantier JC, Damond F, et al. Plasma viral RNA assay in HIV-1 group O infection by real-time PCR. J Virol Methods 2003; 113: 43–9.

[8] Plantier JC, Gueudin M, Damond F, et al. Plasma RNA quantification and HIV-1 divergent strains. J Acquir Immune Defic Syndr 2003; 33: 1–7.

[9] Desire N, Dehee A, Schneider V, et al. Quantification of human immunodeficiency virus type 1 proviral load by a TaqMan real-time PCR assay. J Clin Microbiol 2001; 39: 1303–10.

[10] Zhao Y, Yu M, Miller JW, et al. Quantification of human immunodeficiency virus type 1 proviral DNA by using TaqMan technology. J Clin Microbiol 2002; 40: 675–8.

[11] Saha BK, Tian B, Bucy RP. Quantitation of HIV-1 by real-time PCR with a unique fluorogenic probe. J Virol Methods 2001; 93: 33–42.

[12] Vet JA, Majithia AR, Marras SA, et al. Multiplex detection of four pathogenic retroviruses using molecular beacons. Proc Natl Acad Sci USA 1999; 96: 6394–9.

[13] Hoelscher M, Dowling WE, Sanders-Buell E, et al. Detection of HIV-1 subtypes, recombinants, and dual infections in east Africa by a multi-region hybridization assay. AIDS 2002; 16: 2055–64.

[14] Teo IA, Morlese J, Choi JW, et al. Reliable and reproducible LightCycler qPCR for HIV-1 DNA 2-LTR circles. J Immunol Methods 2002; 270: 109–18.

[15] Yun Z, Fredriksson E, Sonnerborg A. Quantification of human immunodeficiency virus type 1 proviral DNA by the TaqMan real-time PCR assay. J Clin Microbiol. 2002; 40: 3883–4.

[16] Brussel A, Sonigo P. Analysis of early human immunodeficiency virus type 1 DNA synthesis by use of a new sensitive assay for quantifying integrated provirus. J Virol 2003; 77: 10119–24.

[17] Eriksson LE, Leitner T, Wahren B, et al. A multiplex real-time PCR for quantification of HIV-1 DNA and the human albumin gene in CD4+ cells. APMIS 2003; 111: 625–33.

[18] Weber J, Rangel HR, Chakraborty B, et al. A novel TaqMan real-time PCR assay to estimate ex vivo human immunodeficiency virus type 1 fitness in the era of multi-target (pol and env) antiretroviral therapy. J Gen Virol 2003; 84: 2217–28.

[19] Damond F, Descamps D, Farfara I, et al. Quantification of proviral load of human immunodeficiency virus type 2 subtypes A and B using real-time PCR. J Clin Microbiol 2001; 39: 4264–8.

[20] Damond F, Gueudin M, Pueyo S, et al. Plasma RNA viral load in human immunodeficiency virus type 2 subtype A and subtype B infections. J Clin Microbiol 2002; 40: 3654–9.

[21] Schutten M, van den Hoogen B, van der Ende ME, et al. Development of a real-time quantitative RT-PCR for the detection of HIV-2 RNA in plasma. J Virol Methods 2000; 88: 81–7.

[22] Nagai M, Usuku K, Matsumoto W, et al. Analysis of HTLV-I proviral load in 202 HAM/TSP patients and 243 asymptomatic HTLV-I carriers: high proviral load strongly predisposes to HAM/TSP. J Neurovirol 1998; 4: 586–93.

[23] Kamihira S, Dateki N, Sugahara K, et al. Real-time polymerase chain reaction for quantification of HTLV-1 proviral load: application for analyzing aberrant integration of the proviral DNA in adult T-cell leukemia. Int J Hematol 2000; 72: 79–84.

[24] Miley WJ, Suryanarayana K, Manns A, et al. Real-time polymerase chain reaction assay for cell-associated HTLV type I DNA viral load. AIDS Res Hum Retrov 2000; 16: 665–75.

[25] Dehee A, Cesaire R, Desire N, et al. Quantitation of HTLV-I proviral load by a TaqMan real-time PCR assay. J Virol Methods 2002; 102: 37–51.

[26] Kamihira S, Dateki N, Sugahara K, et al. Significance of HTLV-1 proviral load quantification by real-time PCR as a surrogate marker for HTLV-1-infected cell count. Clin Lab Haematol 2003; 25: 111–7.

[27] Estes MC, Sevall JS. Multiplex PCR using real time DNA amplification for the rapid detection and quantitation of HTLV I or II. Mol Cell Probes 2003; 17: 59–68.

[28] Abe A, Inoue K, Tanaka T, et al. Quantitation of hepatitis B virus genomic DNA by real-time detection PCR. J Clin Microbiol 1999; 37: 2899–903.
[29] Ho SK, Yam WC, Leung ET, et al. Rapid quantification of hepatitis B virus DNA by real-time PCR using fluorescent hybridization probes. J Med Microbiol 2003; 52: 397–402.
[30] Loeb KR, Jerome KR, Goddard J, et al. High-throughput quantitative analysis of hepatitis B virus DNA in serum using the TaqMan fluorogenic detection system. Hepatology 2000; 32: 626–9.
[31] Pas SD, Fries E, De Man RA, et al. Development of a quantitative real-time detection assay for hepatitis B virus DNA and comparison with two commercial assays. J Clin Microbiol 2000; 38: 2897–901.
[32] Pas SD, Niesters HG. Detection of HBV DNA using real time analysis. J Clin Virol 2002; 25: 93–4.
[33] Weinberger KM, Wiedenmann E, Bohm S, et al. Sensitive and accurate quantitation of hepatitis B virus DNA using a kinetic fluorescence detection system (TaqMan PCR). J Virol Methods 2000; 85: 75–82.
[34] Chen RW, Piiparinen H, Seppanen M, et al. Real-time PCR for detection and quantitation of hepatitis B virus DNA. J Med Virol 2001; 65: 250–6.
[35] Cane PA, Cook P, Ratcliffe D, et al. Use of real-time PCR and fluorimetry to detect lamivudine resistance-associated mutations in hepatitis B virus. Antimicrob Agents Chemother 1999; 43: 1600–8.
[36] Whalley SA, Brown D, Teo CG, et al. Monitoring the emergence of hepatitis B virus polymerase gene variants during lamivudine therapy using the LightCycler. J Clin Microbiol 2001; 39: 1456–9.
[37] Zhang M, Gong Y, Osiowy C, et al. Rapid detection of hepatitis B virus mutations using real-time PCR and melting curve analysis. Hepatology 2002; 36: 723–8.
[38] Zanella I, Rossini A, Domenighini D, et al. Real-time quantitation of hepatitis B virus (HBV) DNA in tumorous and surrounding tissue from patients with hepatocellular carcinoma. J Med Virol 2002; 68: 494–9.
[39] Kishimoto H, Yoshioka K, Yano M, et al. Real-time detection system for quantitation of hepatitis C virus RNA: a comparison with the other three methods. Hepatol Res 2001; 19: 12–21.
[40] Schroter M, Zollner B, Schafer P, et al. Quantitative detection of hepatitis C virus RNA by light cycler PCR and comparison with two different PCR assays. J Clin Microbiol 2001; 39: 765–8.
[41] Schroter M, Zollner B, Schafer P, et al. Genotyping of hepatitis C virus types 1, 2, 3, and 4 by a one-step LightCycler method using three different pairs of hybridization probes. J Clin Microbiol 2002; 40: 2046–50.
[42] White PA, Pan Y, Freeman AJ, et al. Quantification of hepatitis C virus in human liver and serum samples by using LightCycler reverse transcriptase PCR. J Clin Microbiol 2002; 40: 4346–8.
[43] Kawai S, Yokosuka O, Kanda T, et al. Quantification of hepatitis C virus by TaqMan PCR: comparison with HCV Amplicor Monitor assay. J Med Virol 1999; 58: 121–6.
[44] Mercier B, Burlot L, Ferec C. Simultaneous screening for HBV DNA and HCV RNA genomes in blood donations using a novel TaqMan PCR assay. J Virol Methods 1999; 77: 1–9.
[45] Enomoto M, Nishiguchi S, Shiomi S, et al. Changes in serum levels of hepatitis C virus genotype 1b monitored by real-time quantitative polymerase chain reaction as a predictor of long term response to interferon-alpha treatment. Am J Gastroenterol 2002; 97: 420–6.
[46] Kleiber J, Walter T, Haberhausen G, et al. Performance characteristics of a quantitative, homogeneous TaqMan RT-PCR test for HCV RNA. J Mol Diagn 2000; 2: 158–66.
[47] Mitsunaga S, Fujimura K, Matsumoto C, et al. High-throughput HBV DNA and HCV RNA detection system using a nucleic acid purification robot and real-time detection PCR: its application to analysis of posttransfusion hepatitis. Transfusion 2002; 42: 100–6.
[48] Yang JH, Lai JP, Douglas SD, et al. Real-time RT-PCR for quantitation of hepatitis C virus RNA. J Virol Methods 2002; 102: 119–28.
[49] Bullock GC, Bruns DE, Haverstick DM. Hepatitis C genotype determination by melting curve analysis with a single set of fluorescence resonance energy transfer probes. Clin Chem 2002; 48: 2147–54.

Emerging Biological Threat
G. Berencsi et al. (Eds.)
IOS Press, 2005

Diagnostics Development Research Within the U.S. Biological Defense Research Programme

George V. LUDWIG
US Army Medical Research Institute of Infectious Diseases
Fort Detrick, Maryland, USA

The U.S. Army Medical Research Institute of Infectious Disease's program for development of diagnostic technologies for defense against biological weapons is focused on supporting unit readiness by improving the health of soldiers in the field. The program explores the use of orthogonal technologies to detect multiple agent biomarkers to provide a high level of confidence to the diagnostic questions being asked. Currently, the program research goals are focused in four major areas: assay development, biological target identification, technology assessment, and testing and evaluation. Research conducted over the last seven years has led to the fielding of a large number of real-time nucleic acid detection assays and improved immuno assays capable of detecting all of the major biothreat agents. Future research efforts are focused on integration of technologies into a single device capable of providing high confidence diagnostics to forward deployed troops.

1. Defining the Need

The mission of any country's military is, in very simple terms, to protect and defend the national security of that country. It is therefore, the mission of the military medical system, in relationship to the general mission of the military, to maintain unit readiness and the capability of the fighting force to successfully complete its defensive mission. Any epidemic illness, particularly those caused by biothreat agents or other contagious or highly transmissible diseases have the potential to drastically affect unit readiness and thus the ultimate outcome of a battle or conflict.

Diseases with the potential to cause epidemics are obviously more important than those that occur rarely or only occasionally in at-risk populations. Biological attacks present significant risks to troops because their strategic or even tactical use could have significant effects on unit readiness. Biological agents that traditionally occur in endemic foci and rarely cause explosive outbreaks (anthrax for example) could potentially infect and kill or incapacitate thousands of troops if used as a weapon. It is, therefore, essential that appropriate precautions be taken to reduce the effect (morbidity and mortality) of biological weapons and other, naturally occurring infectious and non-infectious diseases in the military.

In general terms, protecting troops from biological attack and other infectious and non-infectious diseases can be accomplished in three ways: 1) use of personal protective equipment to eliminate or reduce infection and/or transmission of the agents, 2) vaccination or prophylactic pretreatment against specific agents, and 3) early detection and diagnostics to

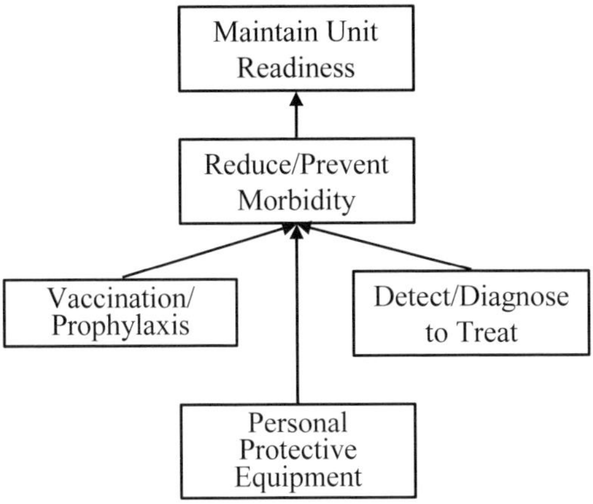

Figure 1. An overview of processes and products used to reduce or prevent the threat of biological weapons on the battlefield with the ultimate goal being to maintain fighting capability of the unit.

facilitate effective treatment before, or early after symptoms begin to appear (Fig. 1). While options one and two have significant importance to the concept of maintaining unit readiness, the need for appropriate diagnostic capabilities will always be necessary, even in the presence of effective vaccines and/or prophylactics. Some portion of the populations will not be protected from disease, either as a result of being missed during the vaccination process or due to incomplete protection induced by the vaccine or prophylactic measures. To be effective, diagnostic capabilities must be rapid, accurate, and reliable. On the surface, these concepts are not unique to the military; however, there are challenges associated with each of these concepts that are military-specific. In addition, each concept has interdependencies with the other concepts that together help form the requirements for a diagnostic system.

The speed with which a diagnosis can be made is critical to the eventual outcome of any single event and ultimately to the health and welfare of fighting units. Definitive diagnoses that require more than 24 hr to accomplish have little positive impact on the prognosis of a patient exposed to most of the important biological agents (Fig. 2). Many biothreat agents can be most effectively treated very early after infection. For example, antibiotic therapy for treating inhalation anthrax is most effective when initiated within 24 hr of exposure [1]. Similarly, antiviral drug therapy for treating poxvirus infections in animals is most effective when administered within the first 24 hr after exposure [2]. By the time a patient becomes acutely or critically ill, treatment is often limited to supportive care. The patient will survive or succumb to infection, based on his or her own ability to fight off infection, and the quality of the supportive care given during the critical phases of the disease. The speed of the diagnosis does not depend solely on the speed of the diagnostic assay or detection system being used. Other factors such as the location of the diagnostic equipment, the efficiency of the sample transport system, and the information management system used to report results all play important roles in ensuring a rapid diagnosis.

While the speed of diagnosis is of primary importance, the accuracy and reliability of the assay system are also critical factors. For the purposes of this discussion, accuracy is defined as the ability of the diagnostic system to correctly identify the infecting etiologic agent. Accuracy encompasses a variety of assay system performance characteristics, including sensitivity and specificity.

Sensitivity and specificity are two assay performance characteristics that are often closely associated. During assay development, specificity is often sacrificed to increase as-

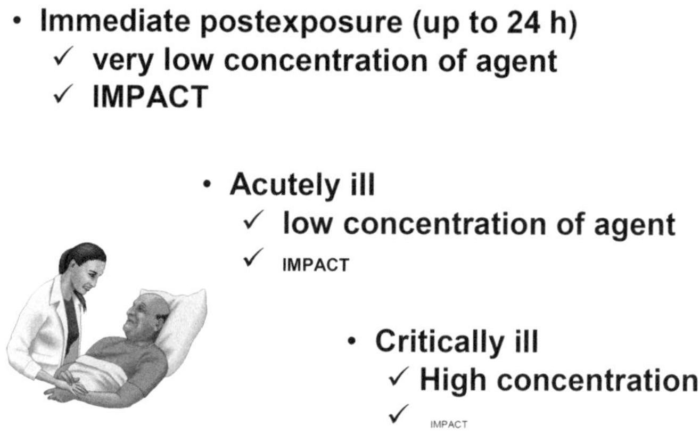

Figure 2. The impact of diagnostics on patient care and ultimately patient prognosis is dependent on the speed with which high confidence diagnostics can be completed.

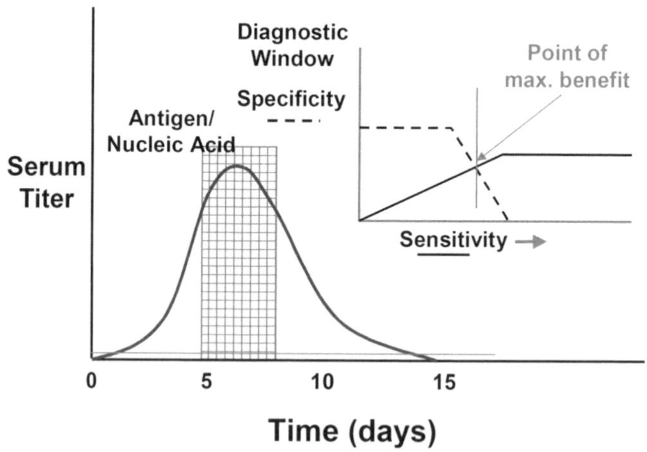

Figure 3. Typical viremia kinetics highlighting the importance of understanding the relationship between assay sensitivity and specificity.

say sensitivity. Similarly, maximizing specificity often comes at the expense of decreased sensitivity. A careful balance between sensitivity and specificity is required and the need for, and importance of, either assay characteristic may vary depending on the goals of the assay being developed.

During the disease process, diagnostic markers appear and disappear at different times. For example, during a classical viral infection, viral particles appear in the blood only during a very narrow period of time and the period of maximum viral titer in the blood is often very short. An assay capable of detecting only the highest levels of viremia will only be an effective diagnostic tool when used during the period of maximal viremia. Increasing the sensitivity of the assay increases the period of time a diagnostic assay can be used to diagnose disease (the diagnostic window). However, assay sensitivity can only be increased so much before the required specificity of the assay decreases below acceptable levels (Fig. 3). The concept of using multiple integrated diagnostic technologies capable of detecting several different biological markers reduces the dependency of a diagnostic system on high sensitivity. This concept for an integrated diagnostic system also addresses the other questions regarding assay accuracy by providing for a system in which multiple assays serve as independent confirmatory tests.

Developing a diagnostics system capable of detecting multiple biomarkers serves several purposes. It increases confidence in the final diagnosis and it helps prevent technological surprise. A diagnostic assay capable of only measuring a single biomarker could conceivably be made unreliable through relatively simple natural or unnatural genetic or biochemical manipulation of the organism being tested. Protein epitopes can be modified such that antibodies no longer bind to the proteins. Diagnostic assays based on these antibodies would no longer function as designed. Similarly, modifying specific gene loci may reduce the effectiveness of, or negate completely nucleic acid detection assays directed toward those loci. However, by incorporating both nucleic acid detection assays directed against multiple genetic targets and antigen-detection assays directed against multiple protein antigens, the likelihood of defeating a given diagnostic system is extremely remote. Of equal importance is the concept that an integrated approach reduces the dependency of the system on sensitivity by incorporation of assays capable of detecting different biomarkers appearing at different times during the infection. For example, a system that could detect early and specific host response markers, agent-specific nucleic acid, agent-specific antigen, and agent-specific antibody could conceivably be used to diagnose infection at any time after exposure. In fact, the use of such a diagnostic system in combination with a detailed understanding of the pathogenesis of the agent could be used to determine the exact stage of a patient's infection. This capability would have far-reaching implications for patient treatment as well as for disease prevention in cases where the probability of disease transmission is high.

These concepts have helped us to formulate an approach to developing an integrated diagnostic system that could meet the U.S. Department of Defense requirements for speed, accuracy, and reliability. As previously discussed, a diagnostic system is more than the diagnostic platform and its component assays. It involves a system of laboratory support that can be deployed to wherever it is needed the most.

The need for a forward-deployed laboratory was first defined during Operation Desert Storm where a rapidly moving force first faced the possible use of biological weapons in addition to the threat of many important naturally occurring infectious diseases. This force required laboratory support that could rapidly identify agents and diagnose disease. The concepts borne out of that deployment were refined during Operation Desert Thunder and subsequent deployments and operational missions into a concept for an integrated battlefield medical environment. For such a system, inputs from bioagent detectors, medical treatment facilities, and environmental recon teams feed into a mobile laboratory capable of field-based confirmatory detection and diagnosis by using an integrated approach. This laboratory possesses reach-back capability to CONUS-based reference laboratories that are capable of providing logistical support, scientific expertise, and final "gold standard" confirmation of results.

2. Diagnostic System

The traditional view of a diagnostic test includes the diagnostic platform, assays, and reagents necessary to make a diagnosis. From a military systems development standpoint, a diagnostic test must incorporate not only all of the components necessary to make a test work, but also the protocols, validation, and testing necessary to ensure that results can be accurately and reliably obtained by the soldiers assigned to this task. More important from a military perspective is the requirement for training and the concepts of operation devel-

Table 1. Important operational requirements for the Joint Biological Agent Identification and Diagnostic System.

Requirement
Simultaneous identification of >= 10 agents
Identify biological agents at relevant concentration
Sample processing <= 20 min
Total assay time <= 25 min
Sensitivity >= 0.98
Specificity >= 0.98
Approved or exempted by the FDA
Upgradeable – equipment/assays
Self-calibrating with failure alert
Protect and preserve samples
Software/communications compatibility
Set up <= 30 min
Visual and audible positive identification signal
Operable using all DoD power systems
Battery operated >= 12 hr
Back-up power capable
Person portable
Day/Night Operations
Onboard data storage

opment required to make the system function in the field. In other words, no diagnostic system will be useful if the pool of available technicians are insufficiently trained in the required laboratory skills and do not understand what to do with system data once the assays have been completed. The latter issues are of paramount importance to the battlefield commander because diagnostic information acquired through the use of an integrated diagnostic system will influence command decisions that may affect the outcome of a battle or conflict.

Given the importance of biothreat agent diagnostic and detection capabilities on the battlefield, a number of requirements have been generated for developing an integrated diagnostic platform (Table 1). Because the ideal diagnostic system is not currently technologically feasible, an acquisition program for developing such a system has been designed in the block improvement model. The first component of the system will be development and deployment of a rapid nucleic acid detection platform capable of accurately detecting and identifying multiple biothreat agents. The second block or first major improvement to be added to the system will be an improved immunodiagnostic system that will add capability for detecting and identifying biological toxins and as a confirmatory and/or preliminary testing platform in support of rapid nucleic acid detection. The last phased improvement to the system will be the complete integration of rapid nucleic acid and antigen detection or other newer technologies to create the integrated common diagnostic platform (ICDP).

3. Development Program Strategic Plan

The operational requirements have been used as a guide for developing a strategic plan for an integrated diagnostic system that is based on four main research areas. These research areas include: 1) assay development, 2) identification of novel biological targets, 3) testing and evaluation, and 4) technology assessment.

Assay development is a diverse research area that includes support of existing or soon to be deployed systems as well as support for new technologies and block improvements that could be incorporated into the ICDP. Assay development includes improvement of existing assays in terms of sensitivity and specificity in addition to development of new assays capable of detecting and identifying new targets that are part of other research efforts.

Research in this part of the strategic plan also includes efforts to expand the repertoire of available assays to increase the number of biothreat and other infectious disease causing agents that can be detected and identified. Assay development also includes research into improvements in sample processing. Sample processing represents a significant challenge to present and future diagnostic systems and includes such concepts as nucleic acid extraction, bioagent and biomarker concentration, and understanding and eliminating matrix effects on detection and identification systems.

Development of new diagnostic assays would not be possible without detailed knowledge of the biology of each biothreat agent of interest. An understanding of the pathogenesis of biothreat agents is critical to determining what tests should be used to assess the kinetics of infection in patients. Studying disease pathogenesis will help to understand the kinetics of the appearance of specific biological markers of infection and will help determine what assays should be developed and used. Similarly, an understanding of the molecular epidemiology and agent-specific genetic profiles will help determine the origin of infecting strains and ensures that diagnostic assays are capable of detecting all strains and geographic isolates of a given organism. An understanding of host- and agent-specific proteomes and transcriptional responses should provide additional insight into identification of new targets useful for diagnosing disease very early in the course of infection.

Assay systems require appropriate testing and evaluation. The concept for such testing at the research and development level does not necessarily include FDA approval. It does, however, include a variety of components that may assist in the approval process should that be desired or required.

Diagnostic assays against biothreat agents will be difficult if not impossible to validate in the traditional manner. This is a result of the relatively low incidence of natural disease caused by these agents and the almost total lack of cases caused by exposure to infectious aerosols. Animal studies using human surrogates will therefore be necessary to generate samples and validation controls necessary for FDA approval. In addition, laboratory studies will be needed to provide sufficient data to determine the suitability of a given assay for further validation and approval. Finally, field-testing the diagnostic system under conditions similar to those expected in the battlefield is required to determine system reliability indices and to help in the development of concept of operation for the system. Again, concepts of operation will be critical to understanding the operational importance of data acquired from the diagnostic system and will ultimately determine how the data are used and what they mean in terms of operational responses to biological attack or infectious disease epidemics.

Finally, new technologies for detecting and identifying biothreat agents and diagnosis of diseases they cause are constantly being developed. The current technological environment has produced a myriad of approaches to medical diagnostics. Such technologies will ultimately be the source of products that meet all of the requirements for a diagnostic system described above. Where appropriate, these technologies must be carefully evaluated and matured in collaboration with government and private partners. Again, the ultimate goal of this particular research and development program is to develop an integrated system that can provide rapid reliable diagnostic testing capabilities. An integrated approach toward diagnostics is new concept to many and may require directed funding and collaboration to reach program goals in a timely fashion.

4. Current Systems

For the last 8 years, USAMRIID has focused research and development activities on several systems capable of detecting and identifying biothreat agents. These systems include

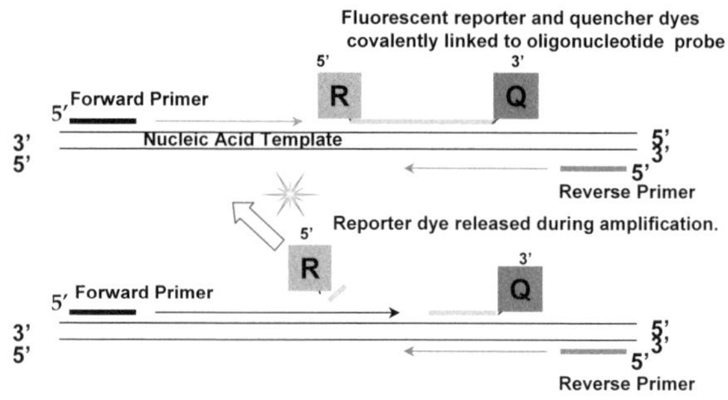

Figure 4. Schematic diagram of the process behind Taqman™ real-time PCR chemistry. R – reporter Dye, Q – quencher dye.

both real-time nucleic acid detection systems and rapid immunodiagnostic systems and form the basis for the concept of integrated diagnostics. The primary candidates for nucleic acid detection systems include the Cepheid Smartcycler™ and the Idaho Technologies RAPID™ [3,4]. Both systems have been hardened to meet military specifications for ruggedness, produce rapid results (25–40 min after sample preparation), are sensitive, and use common fluorescent probe technology. Both systems have distinct advantages that could be exploited in different operational scenarios.

Assays for these two nucleic acid detection systems are available in a variety of fluorescent probe technologies. However, USAMRIID has focused on 5' fluorogenic nuclease assays (Taqman™ assays) that provide increased specificity over several related techniques (Fig. 4). These assays require specific forward and reverse PCR primers as well as a fluorogenically labeled oligonucleotide probe capable of hybridizing to the region of DNA flanked by the two PCR primers. The probe is hydrolyzed during the PCR reaction, leading to the separation of a reporter fluorogenic dye from close proximity to the non-fluorescent quencher dye attached to the opposite end of the probe. This action leads to a concomitant increase in fluorescent signal. As a result, detection occurs in real-time as PCR product is generated. More recent adances in Taqman™ chemistries include the development of DNA minor grove-binding proteins that help to stabilize the probe/template duplex. This capability results in greater flexibility in primer and probe design, improved assay performance, and improved allelic discrimination. Currently, assays capable of detecting over 50 targets from 28 biological agents have been developed and optimized. Many of these assays have been tested against as many as 30,000 samples resulting in the availability of extensive performance data.

As an adjunct to rapid nucleic acid detection assays and platforms, sample-processing procedures have been developed, tested, and evaluated against both environmental samples and medical specimens. Again, processes based mainly on technologies exploiting the capability of nucleic acid to bind to silica and cellulose have been evaluated. These procedures have been tested with over 30,000 environmental samples and many thousands of medical samples.

Several technologies for possible inclusion in an improved immunodiagnostic system are currently being evaluated. These systems include electrochemiluminescence (ECL), time-resolved fluorescence (TRF), flow cytometry, magnetic flux detection, and lateral flow assays. The system with the most promise among those being tested is an ECL system developed by Igen, Inc. ECL assays make use of sandwich assay formats and a chemiluminescent label (ruthenium, Ru). Magnetic beads are coated with capture antibody and, in the presence of biological agent, immune complexes are formed between the agent and the labeled detector antibody.

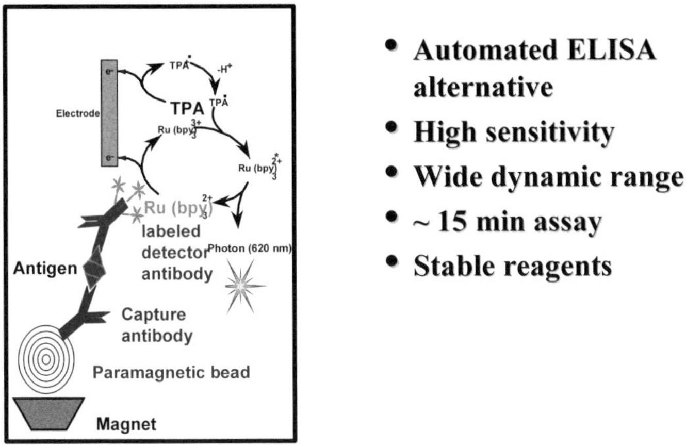

Figure 5. Schematic diagram of the concepts behind Origen™ ECL technology.

The heart of the ECL analyzer is an electrochemical flow cell with a photomultiplier tube (PMT) placed just above the electrode. A magnet positioned just below the electrode captures the magnetic bead-Ru-tagged immune complex and holds it against the electrode. The application of an electric field results in a rapid electron transfer reaction resulting in the production of detectable photons (Fig. 5).

Compared to colorimetric assays like the ELISA, this technology is more sensitive and the instrumentation is simpler. Currently, assays for detecting 14 targets from 10 biothreat agents have been optimized and many of those assays have been extensively tested against environmental and medical samples [3,5].

5. The Future

Current systems meet the immediate needs of the military for rapid and reliable diagnostic capability. However, these systems require extensive logistical support in terms of reagents, repairs, and training. Future systems must be smaller, easier to operate, and require few if any reagents. The challenges for developing and fielding such a future system are many. However, a robust research and development program using expertise from many collaborations involving multiple biological and engineering disciplines will someday be successful in producing a system that meets all of the requirements of the military medical community.

References

[1] Friedlander A.M., Welkos S.L., Pitt M.L., et al. Postexposure prophylaxis against experimental inhalation anthrax. J. Infect. Dis. 1993; 167(5):1239–43.

[2] Roy C.J., Baker R., Washburn K., Bray M. Aerosolized cidofovir is retained in the respiratory tract and protects mice against intranasal cowpox virus challenge. Antimicrob. Agents Chemother. 2003; 47(9): 2933–7.

[3] Henchal E.A., Teska J.D., Ludwig G.V., Shoemaker D.R., Ezzell J.W. Current laboratory methods for biological threat agent identification. Clin. Lab. Med. 2001; 21(3):661–78.

[4] Higgins J.A., Ibrahim M.S., Knauert F.K., et al. Sensitive and rapid identification of biological threat agents. Ann. N.Y. Acad. Sci. 1999; 894:130–48.
[5] Kijek T.M., Rossi C.A., Moss D., Parker R.W., Henchal E.A. Rapid and sensitive immunomagnetic-electrochemiluminescent detection of staphylococcal enterotoxin B. J. Immunol. Methods 2000; 236 (1–2):9–17.

Emerging Biological Threat
G. Berencsi et al. (Eds.)
IOS Press, 2005

New Molecular Targets for Control of *Yersinia Pestis* Infection

Wieslaw SWIETNICKI, Kamal U. SAIKH, Teri KISSNER,
Beverly K. DYAS, Afroz SULTANA and Robert G. ULRICH
Laboratory of Molecular Immunology,
Army Medical Research Institute of Infectious Diseases, Frederick, MD USA

Infrequent infections of domestic animals and humans by *Yersinia pestis*, an endemic bacterial pathogen in many regions of the world, is a result of transmission by the bite of an infected flea and usually terminates with regional lymphadenitis (bubonic plague). However, progression to bacterial septicemia may result in lung colonization, organ failure and death in a high percentage of patients unless infection is controlled by early intervention with antibiotics. Due to the highly virulent nature of infections caused by inhalation of bacterial aerosols *Y. pestis* is listed by federal agencies as a Biodefense Category A pathogen. The protected intracellular bacterial growth of the earliest stage of infection suggest that cytolytic T cells (CTL) and innate immunity are critical to bacterial clearance. Several protein antigens recognized by CTL or antibodies are expressed by *Y. pestis* as components of the type III secretion system. Vaccine or therapeutic strategies targeting this virulence assembly may have the added benefit of providing cross-species protection.

1. Plague

The genetic split of the gram-negative bacillus *Yersinia pestis* from the closely related *Y. pseudotuberculosis*, perhaps 1,500–20,000 years ago [1], coincided with the beginning of periodic and devastating epidemics of bubonic and pneumonic plague occurring throughout most of the recorded historical human world. The cycle of epidemics was broken by improvements in sanitation and the development of antibiotics, although the disease still persists throughout most of the world in isolated human and animal infections. Today there is renewed concern about this ancient scourge of mankind due to the isolation of antibiotic-resistant strains during recent outbreaks and the potential use by bioterrorists. The bacteria are normally transmitted by the bite from an infected flea, which has fed on an infected animal, and results in the bubonic form of plague, an extremely painful infection of draining lymph nodes, usually in the groin or airpits. A small number of bubonic infections may progress to a septicemic state and secondary pneumonic plague, which can be transmitted from person to person by respiratory droplets. This ease of transmission by aerosol, resulting in cases of primary pneumonic plague, increases the potential for amplification of disease to epidemic proportions. The heightened concern about bioterrorism and possibilities of plague re-emergence have renewed interest in understanding *Y. pestis* pathology and potential means of disease intervention.

Strains of *Y. pestis* are classified into one of three original biovars, based on phenotypic differences and epidemiology. The biovar Antiqua, which reduces nitrate and ferments glycerol, is found in Africa and originates from the strain that caused the first pandemic.

Medievalis, which lost the ability to reduce nitrate, is found in Asia and originated from the second pandemic. The most dispersed biovar, Orientalis, does not ferment glycerol and was associated with the third pandemic. Each biovar segregates genetically, based on chromosome location of the insertional element *IS100* [1]. Acquisition of virulence plasmids facilitating flea transmission and additional chromosomal evolution allowed *Y. pestis* to successfully jump from the relatively simple gastrointestinal bacterium lifestyle exemplified by *Y. pseudotuberculosis* to a more complex, highly regulated disease organism. There does not appear to be a simple one or two gene change that is responsible for the more virulent *pestis* phenotype, but rather multiple, perhaps subtle genetic differences. For example, loss of the lipopolysaccharide O-antigen found in *Y. pseudotuberculosis* appears to be essential for function of the plasminogen activator [2], another virulence factor uniquely acquired by *Y. pestis* and encoded on pPCP1. Plague bacteria have evolved to survive or grow in essentially four different environments: dormancy in burrows inhabited by infected rodents, within the flea gut, within phagocytes of mammalian hosts, and finally as an extracellular infection. The pFra-encoded phospholipase D is essential for bacterial survival within the flea [3,4]. A transient and apparently obligate intracellular infection of monocytes and other phagocytic cells [5] occurs during the earliest stage of plague, followed by rapid extracellular expansion of bacteria in lymph nodes and septic disease.

2. Pivotal Role for the Type III Secretion System

Contact of the bacteria with mammalian cells initiates the injection of several virulence factors (Table 1) targeting phagocytic cell function. The type III secretion system (TTSS) responsible for delivery of the Yersinia outer proteins (Yop) virulence factors is also functional in *Y. enterocolitica, Y. pseudotuberculosis*, as well as many other animal and plant pathogenic bacteria, and shares assembly homologs with the flagellum structure. The plasminogen activator assists extracellular growth but is not required for virulence. The precise bacterial factors responsible for supporting extracellular replication and transition from intracellular growth are not established. The TTSS is an important target for therapy and vaccine development (Table 2) because it is essential for pathogenicity, is conserved among several pathogenic bacterial species, and contains components potentially recognized both by CTL and antibodies [6–8]. Three plasmids of *Y. pestis* encode a complex array of virulence factors:

- pMT1, 100 kb, encoding phospholipase D ("murine toxin") required by flea stage for detoxification of blood products, and capsular fraction 1 protein (F1)
- pPCP1, 9.6kb, encoding plasminogen activator serine protease (Pla) pCD1, 70.3 kb, encoding anti-host proteins (LcrV, Yops), their
- chaperones and other proteins contributing to the TTSS.

Proteins of the TTSS are coordinately assembled by cognate chaperones and regulated by genetic elements triggered by environmental factors such as divalent cation concentrations and temperature. Insertion of most or perhaps all virulence effector proteins into the mammalian host cytosol is accomplished by delivery through an injectisome spanning both bacterial membranes. The injectisomes of *Shigella flexneri* [9,10] and *Salmonella typhimurium* [11,12] have been isolated and studied by electron microscopy. In *Shigella*, the structure is composed of a 10 x 60-nm external needle-like complex, a 30-nm (in diameter) transmembrane domain and, in *Shigella*, a 45 x 25 nm cytoplasmic "bulb" forming the base of the complex. The needle is composed primarily of the MxiH subunits in *Shigella* and PrgI subunits in *Salmonella*. In *Y. enterocolica* and *Y. pestis*, YscF forms the needle protein.

Table 1. Proteins[1] Encoded by pCD1.

ORF[2]	Position	(aa residues)	Description or homolog
1	87–1109	340	Transposase
2	1109–1888	259	Transposase
3	1939–2343	134	Transposase
4	2379–2645	88	*lcrS* (*Y. pseudotuberculosis*)
5	3193–3540	115	*lcrQ* (*Y. pseudotuberculosis*) and *yscM* (*Y. enterocolitica*)
6	3765–4430	221	*YscL* (*Y. enterocolitica*)
7	4376–5005	209	*YscK* (*Y. enterocolitica*)
8	5005–5739	244	*YscJ* (*Y. enterocolitica*)
9	5746–6093	115	*ycsI* (*Y. enterocolitica*) and *lcrO* (*Y. pseudotuberculosis*)
10	6094–6591	165	*yscH* (*Y. enterocolitica*) and *lcrP* (*Y. pseudotuberculosis*)
11	6588–6935	115	*YscG* (*Y. enterocolitica*)
12	6937–7200	87	*YscF* (*Y. enterocolitica*)
13	7201–7401	66	*YscE* (*Y. enterocolitica*)
14	7398–8657	419	*YscD* (*Y. enterocolitica*)
15	8654–10477	607	*YscC* (*Y. enterocolitica*)
16a	10483–10896	137	*YscB* (*Y. enterocolitica*)
16b	11121–11220	32	*YscA* (*Y. enterocolitica*)
17	11299–12114	271	*lcrF* (*virF*) transcription factor
18	12238–12633	131	*VirG* (*Y. enterocolitica*)
19	13209–14273	354	*YscU* (*Y. enterocolitica*)
20	14273–15058	261	*YscT* (*Y. pseudotuberculosis*)
21	15055–15321	87	*YscS* (*Y. pseudotuberculosis*)
22	15323–15976	217	*yscR* (*Y. pestis*)
23	15973–16896	307	*yscQ* (*Y. pestis*)
24	16893–18260	455	*yscP* (*Y. pestis*)
25	18260–18724	154	*yscO* (*Y. pestis*)
26	18721–20040	439	*YscN* (Yop secretion ATPase)
27	20238–21119	293	*YopN* (*Y. pseudotuberculosis*)
28	21100–21378	92	*Y. pseudotuberculosis* hypothetical protein
29	21365–21736	123	*Y. pseudotuberculosis* and *Y. enterocolitica* hypothetical protein
30	21733–22101	122	*Y. enterocolitica* hypothetical protein
31	22098–22442	114	*Y. enterocolitica* hypothetical protein
32	22429–24543	704	*lcrD* (*Y. pseudotuberculosis* and *Y. enterocolitica*)
33	24540–24980	144	*lcrR* (*Y. pestis*)
34	25022–25309	95	*lcrG* (*Y. pestis*)
35	25311–26291	326	*lcrV* (V antigen)
36	26304–26810	168	*lcrH* (*sycD*) (YopB and YopD chaperones)
37	26788–27993	401	*yopB* (*Y. pseudotuberculosis* and *Y. enterocolitica*)
38	28012–28932	306	*yopD* (*Y. pseudotuberculosis* and *Y. enterocolitica*)
39	29345–29512	55	*Y. pestis* hypothetical protein
40	29778–30038	87	*Y. pseudotuberculosis* transposase
41	30873–32102	409	*yopM* (*Y. pestis*)
42	32145–32444	99	Transposase
43	34860–35828	322	*Y. enterocolitica* hypothetical protein
44	36328–36876	182	*yopK* (*Y. pseudotuberculosis*) and *yopQ* (*Y. enterocolitica*)
47	38624–39016	130	Transposase
48	40080–41288	402	*sopA* (*E. coli*)
49	41417–42250	277	*sopB* (*E. coli*)
50	44186–44845	219	*yopE* (*Y. pestis*)
51	45039–45431	130	*sycE* (YopE chaperone)
52	45494–46123	309	Transposase
53	46241–47413	390	Transposase
54	47413–47844	143	Transposase
55	48188–48613	141	*sycH* (YopH chaperone) (*Y. pseudotuberculosis*, *Y.enterocolitica*)
56	49594–49860	88	Transposase
57	50911–51462	183	Transposon gamma-delta resolvase and *tnpR* (*Y. enterocolitica*)
58	51626–53941	771	Transposase and *tnpA* (*Y. enterocolitica*)
59	53938–54318	206	Transposase
62	56488–56297	63	Hypothetical protein (*Y. enterocolitica*)
63	56928–57344	138	DNA helicase I (*E. coli* plasmid F)
64	58681–58929	82	Endonuclease (*Y. enterocolitica*)
65	59067–59321	84	*repB* (replication protein)
66	59618–60496	292	*repA* (replication protein)
67	63100–65298	732	*ypkA* (protein kinase) (*Y. pseudotuberculosis*, *Y. enterocolitica*)
68	65694–66557	287	*yopJ* (*Y. enterocolitica*)
69	67146–67649	167	Transposase
70	68243–69649	468	*yopH* (Tyr-phosphatase)(*Y. pseudotuberculosis*, *Y. enterocolitica*)
71	70502–70161	113	Transposase (partial)

[1]Adapted from: Hu *et al*., J. Bact. 180:5192–5202 (1998); [2]Plasmid open-reading frame.

Table 2. Key Features of the Y. pestis Type III Secretion Apparatus[1].

- Coordinately regulated low-Ca^{++} response stimulon of pCD1
- LcrF regulates maximum expression of LcrV and Yops at 37C: requires cell contact or Ca^{++} depletion
- LcrV and Yops secreted without processing
- Yop transport requires transient complexes with specific chaperones (Syc); delivered by injectisome
- YopH is a protein tyrosine phosphatase, dephosphorylating focal adhesion components
- YpkA/YopO, is serine/threonine kinase, disrupts cytoskeleton; unknown host targets
- YopE is a selective activator of mammalian Rho GTPases, disrupts cytoskeleton
- YopT is cysteine protease cleaving Rho GTPases (RhoA, Rac, Cdc42); disrupts actin microfilaments
- YopJ interferes with intracellular signaling cascades; cysteine protease?
- YopM is transported to cell nuclei; cellular target unknown
- LcrV is required for induction of low calcium response; IL-10 mediated suppression of TNF-α, IFN-γ via TLR2 and CD14

[1]Summary of data from several published reports.

Polymerization of subunits into a needle-like structure was hypothesized to lead to puncturing of mammalian cell membranes [13], but the validity of this observation was also questioned [11]. The needle length in *Yersinia enterocolica* is controlled by YscP [64], a protein found only in *Yersiniae ssp.* A central 2–3 nm channel is present in the needle of *Shigella flexneri*, a diameter apparently only large enough to accommodate partially unfolded Yops [14]. In *Salmonella typhimurium*, the putative order of subunit assembly into the needle-like structure of the injectisome [15] and also the flagellum [16] was proposed, based on genetic analysis. A common evolutionary history for portions of the TTSS export machinery and the basal flagellum assembly process [16] was suggested, based on the similarity in proposed function and sequence between the corresponding systems.

Many of the *Yersiniae* proteins have direct homologues in *Salmonella*. For example, YscF is a homologue of the needle assembly proteins in other bacteria known to use TTSS: PrgI from *Salmonella typhimurium*, MxiH from *Shigella flexneri*, YscF from *Yersinia enterocolitica*, PscF from *Pseudomonas aeruginosa*, and EscF from enteropathogenic *Escherichia coli* [17]. YscC protein was found to form a 20 nm-diameter ring with an opening on the surface of *Yersinia enterocolica* [18]. However, there appears to be no obvious *Y. pestis* homologue of the *Salmonella* InvJ protein, a regulator of transport through the TTSS needle assembly. YscN is a 47.8-kDa protein related to the catalytic subunits of FoF1 [19], and related ATPases, and FliI, a protein from *Salmonella* involved in the assembly of flagellar export apparatus [16]. In *Shigella flexneri*, the *spa47* protein is required for both flagellar and TTSS functions [20], and is a homolog of the *Y. pestis* YscN protein. Most bacterial species harboring a functional TTSS also express an associated □ (protein dimer) /□ (protein dimer) subunit of ATP synthase (W. Swietnicki, unpublished data). Previous studies demonstrated that H^+-driven and Na^+-driven motors drive flagellar rotation [21,22]. For example, the *Escherichia coli* flagellum motor is driven by a proton gradient [23], whereas, the *Vibrio cholera* motor is postulated to be driven by sodium ion fluxes [24]. It is possible, but unconfirmed, that transport and assembly of TTSS in *Yersiniae* may be powered by similar mechanisms. Alternatively, recent data from a study of the Caf1M crystal structure suggested that the chaperone preserves the subunit folding energy necessary to drive F1 fiber assembly [25].

3. Critical TTSS Protein-Protein Interactions

Protein-protein interactions within the TTSS assembly have been proposed for some protein pairs. Most of the studies were conducted using deletion/replacements strategies target-

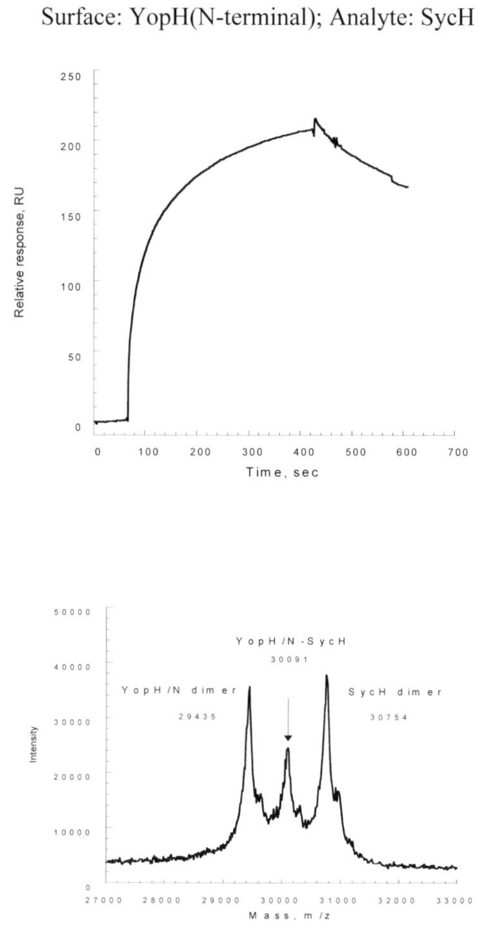

Figure 1. A representative surface plasmon resonance sensorgram (top panel) and a corresponding mass spectrogram (MALDI-TOF) (bottom panel) for interaction between SycH and YopH N-terminal fragment. All measurements were done at 37° C.

ing a single protein or by the yeast two-hybrid method [26]. To provide more precise data, we are employing an alternative strategy to directly measure binding events. A major objective is to construct an interaction matrix for the all of the known TTSS proteins. Preliminary data suggests that some components may have several interaction partners. In the first stage, the proteins are screened by surface plasmon resonance and the potential interacting pairs are then reexamined by mass spectrometry. An example of this analysis is shown in Fig. 1 for the amino-terminal domain of YopH and the chaperone sycH. This approach allows direct measurement of the strength and kinetics of interactions, rather than the indirect data obtained from genetic screens. We anticipate that one or more of these critical interactions will result in new targets for inhibitor drug design.

4. Therapeutic Targets

Potential host cell targets of the tyrosine phosphatase (PTPase) activity of YopH, demonstrated by a variety of means, are the focal adhesion proteins Crk-associated substrate (p130Cas) and focal adhesion kinase (FAK) in epithelial cells, and p130Cas and Fyn-binding protein (Fyb) in macrophages. Bacteria do not appear to contain phosphorylated proteins, suggesting that the enzymatic activity of YopH has evolved solely for diverting host immunity. Because inhibitors of YopH may be valuable therapeutic agents, PTPase inhibitors have been examined for the potential therapeutic benefit of blocking *Y. pestis* infection. In one study, 26 YopH inhibitors with IC_{50} values below 100 μM were identified from a chemical diversity library of 720 structurally diverse carboxylic acids [27]. The most potent and specific YopH inhibitor was aurintricarboxylic acid (ATA), which exhibits a *K*i value of 5 nM for YopH and displays 6 to 120-fold selectivity in favor of YopH against a panel of mammalian PTPs. The inhibitor ATA appeared to block the inhibitory activity of YopH on cellular tyrosine phosphorylation, ERK1/2 activity, and interleukin-2 transcriptional activity, thus restoring normal cell function. Inhibitory potencies were studied with a variety of phosphotyrosyl (pTyr) mimetics against the human PTP1B enzyme [28] by displaying them in the EGFR-derived hexapeptide sequence, 'Ac-Asp-Ala-Asp-Glu-Xxx-Leu-amide', where Xxx=pTyr mimetic. The poor inhibitory potencies of certain of these pTyr mimetics were attributed to restricted orientation within the PTP1B catalytic pocket incurred by extensive peripheral interaction of the hexapeptide platform. Utilizing the smaller tripeptide platform, 'Fmoc-Glu-Xxx-Leu-amide' it was demonstrated that several of the low affinity hexapeptide-expressed pTyr mimetics exhibit high PTP1B affinity within the context of the tripeptide platform. Of particular note, the mono-anionic 4- (carboxydifluoromethyl)Phe residue exhibits affinity equivalent to the di-anionic F2Pmp residue, which had previously been among the most potent PTP-binding motifs. Against YopH, it was found that all tripeptides having Glu residues with an unprotected side chain carboxyl were inactive. Alternatively, in their Glu-OBn ester forms, several of the tripeptides exhibited good YopH inhibitory activity, with the mono-anionic peptide, Fmoc-Glu(OBn)-Xxx-Leu-amide, where Xxx=4 (carboxymethyloxy)Phe providing an IC_{50} value of 2.8 mM.

5. Novel Vaccine Targets

All stages of disease progression are potential targets of the host immune response. Phagocytosis by neutrophils and macrophage, inflammatory cytokine responses triggered by pathogen-associated molecular patterns (PAMPs) through Toll-like receptors and additional sentinel proteins, and natural killer (NK) cell activity, are all innate immune responses functioning in the naïve host to clear invading pathogens. Circulating antibodies from prior immunity or specific vaccination will presumably aid in rapid clearance of bacteria from the stage of flea injection. Phagocytosis and antibody-mediated immunity have been extensively studied. In contrast, little is known about the role of other innate immune mechanisms as well as adaptive T-cell immunity in clearance of plague bacteria. Due to the protected intracellular nature of the initial *Y. pestis* infection and the unusual burden placed on the immune system by pneumonic infection, cytolytic T cells (CTL) and perhaps an enhanced innate response will be important for sterilizing immunity. Manufacturing of the US-licensed, formaldehyde-killed, whole bacilli vaccine was discontinued in 1999 and is no longer available. There is considerable interest in understanding the human immune response to plague and in the discovery of new vaccines. Vaccines against *Y. pestis* infection have primarily focused on the fraction 1 capsular (F1) and V (LcrV) antigens and antibody-mediated immune protection [29,30]. The F1 polypeptide, encoded on pMT1, is synthesized in large quantities by the pathogen, and is reported to confer anti-phagocyte properties on *Y. pestis* by interfering with complement-mediated opsonization. LcrV is a secreted protein encoded by pCD1, and appears to be multifunctional protein required for regulation of Yop production and translocation into host cells.

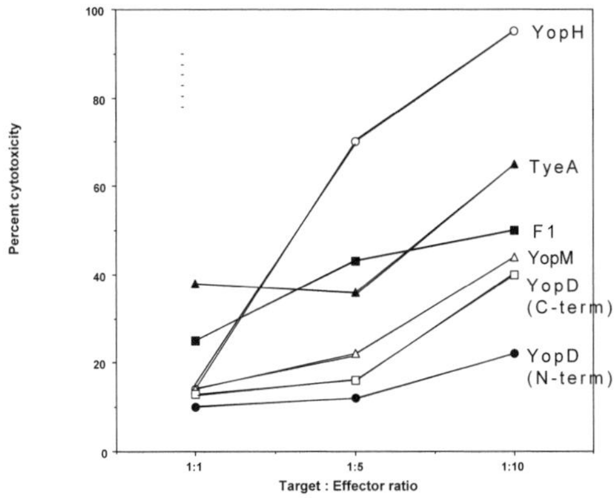

Figure 2. Human cytolytic T-cell recognition of TTSS antigens from monocytes infected with Y. pestis. Primary cultures of human monocytes were infected with Y. pestis (pgm-CO92) and cultured with naïve human T cells. HLA-A2 homozygous B lymphoblastoid cell target cells were pulsed with recombinant TTSS proteins as indicated and target cell lysis in the presence of T cells was measured by ATP cell content. Nonspecific lysis subtracted.

Besides antibody responses, cell mediated immune responses are reported to have a role in protective immunity with *Y. enterocolitica* infection in mice [31]. However, there appears to be a research gap in understanding of the role of cell-mediated immunity in the immune response to *Y. pestis* protective antigens. We examined *in vitro* human T-cell responses in primary culture to recombinant Yop proteins. Monocytes or dendritic cells (DC) infected with *Y. pestis* were used as target cells for stimulating to T cells, and purified recombinant proteins of *Y. pestis* were used as antigens to examine specific T-cell recognition. Several TTSS proteins were recognized by naïve T-cells in primary culture, as indicated in the example shown in Fig. 2. TTSS injected proteins, such as YopH, and the bacterial surface filament component F1 were both recognized as processed antigens presented by infected monocytes. Collectively, these results demonstrate that both helper and cytotoxic T-lymphocyte responses can be detected against some components of the TTSS and indicate the utility in exploring new vaccine targets. The CTL responses may be due to the intracellular expression of the recognized antigens. Alternatively, induction of cytotoxic T-cell immunity sometimes requires the phagocytosis of pathogens or apoptotic bodies, virus-infected or dead tumor cells by professional antigen presenting cells such as DC [33]. Peptides derived from phagocytosed antigens are presented to $CD8^+$ T lymphocytes on major histocompatibility complex (MHC) class I molecules, a process called "cross-presentation". Phagocytosed plague bacterium [5] may allow cross presentation of apoptotic bodies [34] containing TTSS peptides. As an example, YopB and YopD comprise the portion of the injectisome which is thought to insert into the mammalian host membrane. Although YopB and YopD are not directly involved in apoptosis induction, they facilitate the translocation of Yop effector proteins and are hypothesized to associate tightly with cell membranes. Antibody responses to YopD were detected in mice [8] and YopD was shown to protect mice against lethal *Y. pestis* challenge [30]. Soluble YopD presented by DC and monocytes cultures is also recognized by human CTL (unpublished observations).

6. Summary

With the re-emergence of *Yersinia pestis* as a potential threat to the public and deployed military, there is a clear need for new medical approaches for disease intervention. Technical advancements and the availability of whole genome sequences have facilitated the identification of several new molecular targets for developing vaccines and therapeutic drugs. Future study of the type III secretion system, common to several important pathogens, will provide many candidate proteins and novel biochemical pathways to test for critical bacterial targets.

References

[1] Achtman M., Zurthe K., Morelli G., Torrea G., Gulyoule A. and Carniel E. 1999. Proc. Natl. Acad. Sci, USA 96:14043–14048.
[2] Kukkonen M., Suomalainen M., Kyllonen P., Lahteenmaki K., Lang H., Virkola R., Helander I.M., Holst O., Korhonen T.K. 2004. Mol. Microbiol. 51:215–25.
[3] Hinnesbusch B.J., Perry R.D., and Schwan T.G. 1996. Science 273:367–370.
[4] Hinnebusch B.J., Rudolph A.E., Cherepanov P., Dixon J.E., Schwan T.G. and Forsberg A. 2002. Science 296:733–735.
[5] Straley S.C., Harmon P.A. 1984. Infect. Immun. 45:649–54.
[6] Brandler P., Saikh K.U., Heath D., Friedlander A., and Ulrich R.G. 1998. J Immunol 161:4195–4200.
[7] Starnbach M.N., and Bevan M.J. 1994. J. Immunol. 153:1603–12.
[8] Benner G.E., Andrews G.P., Byrne W.R., Strachan S.D., Sample A.K., Heath D.G., and Friedlander A.M. 1999. Infect. Immun. 67:1922–8.
[9] Tamano K., Aizawa S.-I., Katayama E., Nonaka T., Imajoh-Obi S., Kuwae A., Nagai S., and Sasakawa C. (2000). EMBO J. 19:3876–3887.
[10] Blocker A., Jouihri N., Larquet E., Gounon P., Ebel F., Parsot C., Sansonetti P., and Allaoui A. (2001). Mol. Microbiol. 39:652–63.
[11] Marenne M.-N., Journet L., Mota L.J. and Cornelis G.R. (2003). Micr. Pathog. 35:243–258.
[12] Kimbrough T.G., and Miller S.I. (2000). Proc. Natl. Acad. Sci. USA. 97:11008–11013.
[13] Hoiczyk E. and Blobel G. (2001). Proc. Natl. Acad. Sci. USA. 98:4669–4674. Journet L., Agrain C. Broz P., and Cornelis G.R. (2003). Science 302:1757–1760.
[14] Stebbins C.E., and Galan J.E. (2001). Nature 414, 77–81.
[15] Sukhan A., Kubori T., Wilson J., and Galan J.E. (2001). J. Bacteriol. 183:1159–67.
[16] Macnab R.M. (2003). Annu. Rev. Microbiol. 57:77–100.
[17] Kubori T., Sukhan A., Aizawa S.I., and Galan J.E. (2000). Proc. Natl. Acad. Sci. USA. 97:10225–30.
[18] Koster M., Bitter W., de Cock H., Allaoui A., Cornelis G.R., and Tommassen, (1997). J. Mol Microbiol. 26:789–97.
[19] Woestyn S., Allaoui A., Wattiau P., Cornelis G.R. 1994. YscN, the putative energizer of the Yersinia Yop secretion machinery. J Bacteriol. 176:1561–9.
[20] Tamano K., Aizawa S.-I., Katayama E., Nonaka T., Imajoh-Obi S., Kuwae A., Nagai S., and Sasakawa C. (2000). EMBO J. 19:3876–3887.
[21] Blair D.F. (1995). Annu. Rev. Microbiol. 49:489–522.
[22] Imae Y., and Atsumi T. (1989). J. Bioenerg. Biomembr. 21:705–716.
[23] Gabel C.V., and Berg H.C. (2003). Proc. Natl. Acad. Sci. USA. 100:8748–9751.
[24] Kojima S., Yamamoto K., Kawagishi I., and Homma M. (1999). J. Bacteriol. 181:1927–1930.
[25] Zavialov A.V., Berglund J., Pudney A.F., Fooks L.J., Ibraham T.M., MacIntyre S., and Knight S.D. 2003. Cell 113: 587–596.
[26] Jackson M.W., Plano G.V. 2000. Interactions between type III secretion apparatus components from Yersinia pestis detected using the yeast two-hybrid system. FEMS Microbiol Lett. 186(1):85–90.
[27] Liang F., Huang Z., Lee S.Y., Liang J., Ivanov M.I., Alonso A., Bliska J.B., Lawrence D.S., Mustelin T., Zhang Z.Y. 2003. Aurintricarboxylic acid blocks in vitro and in vivo activity of YopH, an essential virulent factor of Yersinia pestis, the agent of plague. J Biol Chem. 278:41734–41.
[28] Lee K., Gao Ya., Yao Z.-Y., Phan J., Wu L., Liang J., Waugh D.S., Zhang Z.-Y., and Burke T.R. Jr., 2003. Tripeptide inhibitors of yersinia protein-tyrosine phosphatase. Bioorg. Med. Chem. Lett. 13, 2577–2581.
[29] Williamson E.D. 2001. J. Appl. Microbiol. 91:606–608.
[30] Heath D.G., Anderson G.W. Jr., Mauro J.M., Welkos S.L., Andrews G.P., Adamovicz J., Friedlander A.M. Vaccine. 1998. 16:1131–7.
[31] Autenrieth I.B., Vogel U., Preger S., Heymer B., and Heesemann J. 1993. Infect. Immun.61:2585–95.
[32] Harshyne L.A., Watkins S.C., Gambotto A., and Barratt-Boyes S.M. 2001. J Immunol. 166:3717–23.
[33] Mills S.D., Boland A., Sory M.P., van der Smissen P., Kerbourch C., Finlay B.B., and Cornelis G.R. 1997. Proc. Natl. Acad. Sci. USA. 94:12638–43.

Emerging Biological Threat
G. Berencsi et al. (Eds.)
IOS Press, 2005

Enhanced Concentration and Extraction of Bacillus Anthracis DNA from Whole Blood

Matt EWERT, MS, CT (ASCP)
Center for Biological Defense, USF

Previous acts of biological terrorism involving release of *B. anthracis* spores within the United States logically prompted blood testing for the clinical detection of anthrax [1–3]. The mediastinitis that characterizes an initial course of inhalational anthrax is associated with a rapidly progressive bacteremia once efferent lymphatics become highly laden with organisms [2,4]. Therefore, PCR testing of blood should provide a rapid method for detection of anthrax in bacteremic patients [3].

The earliest indication suggesting a diagnosis of inhalation anthrax for this nation's first victim of a *B. anthracis* related act of biological terrorism was hemorrhagic spinal fluid with gram positive bacilli [5]. Culture and PCR were essential to confirm the suspected disease state. In this case however, concentration of pathogen for sensitivity enhancement was completely unnecessary. Other diagnostic efforts however may be supplemented with PCR findings that can be provided early in the course of a disease process. Although clinical symptoms that would prompt testing for biological threat agents may be associated with terminal septicemia, the hematogenous spread of vegetative *B. anthracis* from mediastinal and peribronchial lymph nodes to other organ sites may confound diagnostic attempts by mimicking various disease states [6].

It is the period of initial hematogenous spread that is most appropriate for enhanced sensitivity testing. The potentially disproportionate pediatric casualties expected to result from a biological weapons release may as well require sensitive and timely diagnostic capabilities that guide targeted interventions [7,8].

Levels of bacteremia associated with common bacterial pathogens in adults and children may be less than 1 CFU per ml [9,10] (10, 18, 28). The use of a low volume of blood for a blood culture is known to reduce the detection of bacteremia in adults [11] (13). Some previously described methods for extracting biological threat agent DNA from whole blood use a sample of 1.0 ml or less [12,13] (11, 15, 25). Moreover, commercially available manual nucleic acid extraction kits typically specify as little as 0.2 ml of blood for a single sample-processing event. Clearly, PCR methods for the detection of bloodborne pathogens will have increased sensitivity if blood volumes > 5.0 ml can be processed without co-extracting the entire compliment of human DNA.

Blood element lysis using detergents or osmotic shock, enzymatic action of DNase I or streptokinase followed by bacterial concentration via centrifugation or filtration have been explored in various combinations [10,12–19]. We describe an enzyme/detergent cocktail that selectively solubilizes human blood elements while preserving *B. anthracis*. This approach allows *B. anthracis* to be concentrated from relatively large volumes (>5 ml) of blood via centrifugation or filtration and thereby may increase the sensitivity of PCR testing for this organism. Low blood matrix associated particulate loads contributed by treatment with the enzyme/detergent cocktail allowed rapid filtration (20 sec) and proteinase K digestion (10 min) during nucleic acid extraction.

Samples of human blood were collected in EDTA tubes, seeded with *B. anthracis,* and then treated with our enzyme/detergent cocktail. Following treatment with this cocktail, the bacteria were concentrated by either centrifugation or filtration methods. Nucleic acid was then extracted from these bacterial concentrates using a modified MagNa Pure LC DNA isolation Kit III. PCR analysis was conducted using a LightCycler and the LightCycler *Bacillus anthracis* Detection Kit described elsewhere [20]. The sensitivity of this method was 1–5 CFU of *B. anthracis* per 6-ml of whole blood.

This new method provides a simplified approach to concentrating bacteria from large blood samples by minimizing sample transfer steps. Also, by using mild detergent conditions and enzymes that specifically target the proteins and DNA that comprise the structural composition of blood sample matrices, the method may be compatible with a broad spectrum of bacterial agents. The chemical and biochemical environment produced by the described detergent and enzyme cocktail has potential to be useful for concentrating a broad spectrum of bacteria from blood hence allowing one DNA extract to serve a panel of PCR based tests.

References

[1] Centers for Disease Control and Prevention (CDC). 2001. Investigation of bioterrorism related anthrax and interim guidelines for clinical evaluation of persons with possible anthrax. MMWR. 50:941–948.

[2] Lincoln, R.E., D.R. Hodges, F. Klein, B.G. Mahlandt, W.I. Jones, Jr., B.W. Haines, M.A. Rhian, and J.S. Walker. 1965. Role of lymphatics in the pathogenesis of anthrax. J. Infect. Dis. 115:481–494.

[3] Mina, B., J.P. Dym, F. Kuepper, R. Tso, C. Arrastia, I. Kaplounova, H. Faraj, A. Kwapniewski, C.M. Krol, M. Grosser, J. Glick, S. Fochios, A. Remolina, L. Vasovic, J. Moses, T. Robin, M. DeVita, and M.L. Tapper. 2002. Fatal inhalational anthrax with unknown source of exposure in a 61-year-old woman in New York City. JAMA. 287:858–862.

[4] Ross, J.M. 1957. The pathogenesis of anthrax following the administration of spores by the respiratory route. J. Pathol. Bacteriol. 73:485–494.

[5] Bush, L.M., B.H. Abrams, A. Beall and C.C. Johnson. 2001. Index case of fatal inhalational anthrax due to bioterrorism in the United States. N. Eng. J. Med. 345:1607–1610.

[6] Quintiliani, R. Jr., R. Quintiliani. 2002. Fatal case of inhalational anthrax mimicking intra-abdominal sepsis. Conn. Med. 66:261–267.

[7] Committee on Environmental Health and Committee on Infectious Diseases. 2000. Chemical-Biological Terrorism and Its Impact on Children: A Subject Review. Pediatrics. 105:662–670.

[8] Patt, H.A., R.D. Feigin. 2002. Diagnosis and management of suspected cases of bioterrorism: a pediatric perspective. Pediatrics. 109:685–692.

[9] Kellogg, J.A., J.P. Manzella, and D.A. Bankert. 2000. Frequency of low level bacteremia in children from birth to fifteen years of age. J. Clin. Microbiol. 38:2181–2185.

[10] Reimer, L.G., M.L. Wilson, M.P. Weinstein. 1997. Update on detection of bacteremia and fungemia. Clin. Microbiol. Rev. 10:444–465.

[11] Mermel, L.A., and D.G. Maki. Detection of bacteremia in adults: consequences of culturing an inadequate volume of blood. 1993. Ann. Intern. Med. 119:270–272.

[12] Leal-Klevezas, D.S., I.O. Martinez-Vazquez, A. Lopez-Merino, and J.P. Martinez-Soriano. 1995. Single-step PCR for detection of Brucella spp. from blood and milk of infected animals.

[13] Morata, P., M.I. Queipo-Ortuño, J.M. Reguera, M.A. García-Ordoñez, A. Cárdenas, and J.D. Colmenero. 2003. Development and evaluation of a PCR-enzyme-linked immunosorbent assay for diagnosis of human brucellosis. J. Clin. Microbiol. 41:144–148.

[14] Bernhardt, M., D.R. Pennell, L.S. Almer, and R.F. Schell. 1991. Detection of bacteria in blood by centrifugation and filtration. 29:422–425.

[15] Dorn, G.L., and K. Smith. 1978. New centrifugation blood culture device. J. Clin. Microbiol. 7:52–54.

[16] Dorn, G.L., G.A. Land, and G. E. Wilson. 1979. Improved blood culture techniques based on centrifugation: clinical evaluation. J. Clin. Microbiol. 9:391–396.

[17] Gamboa, F., J.M. Manterola, J. Lonca, L. Matas, B. Vinado, M. Gimenez, P.J. Cardona, E. Padilla, and V. Ausina. 1997. Detection and identification of mycobacteria by amplification of RNA and DNA in pretreated blood and bone marrow aspirates by a simple lysis method. J. Clin. Microbiol. 35:2124–2128.

[18] Hoffman, S., D.C. Edman, N.H. Punjabi, M. Lesmana, A. Cholid, S. Sundah, and J. Harahap. 1986. Bone marrow aspirate culture superior to streptokinase clot culture and 8 ml 1:10 blood-to-broth ratio blood culture for diagnosis of typhoid fever. Am. J. Trop. Med. Hyg. 35:836–839.

[19] Zhaoqin, M., C. Jiankui, Y. Xiuyun, Z. Lixue. 1998. Polymerase chain reaction for the diagnosis of candidemia. Bull. Acad. Mil. Med. Sci. 22:218–220.

[20] Bell, C.A., J.R. Uhl, T.L. Hadfield, J.C. David, R.F. Meyer, T.F. Smith, and F.R. Cockerill III. 2002. Detection of Bacillus anthracis DNA by LightCycler PCR. J. Clin. Microbiol. 40:2897–2902.

Emerging Biological Threat
G. Berencsi et al. (Eds.)
IOS Press, 2005

Biological Toxins and Super-Antigens as an Emerging Biological Threat

Akbar S. KHAN* and James VALDES
Molecular Engineering Team, U.S Army Soldier and Biological Chemical Command, 5183 Blackhawk Road, Edgewood CB Center, APG, MD 21010-5424
akbar.khan1@us.army.mil

The current revolution in biology especially genomics and proteomics, has identified genes encoding for new biological toxins and super-antigens. There is growing concern within both scientific defense and intelligence communities that this constitutes a serious potential for misuse as offensive biological weapons. Currently, sequences of close to 50 microbial genomes have been completed and the sequences of more than 100 genomes should be completed within the next 2 to 5 years. These sequences will encode a collection of >200,000 predicted coding sequences which will code for important functional proteins, as well as potential new biological toxins and super-antigens. Completed sequences of microbial genomes provide an excellent source to study the physiology and evolution of microbial species and expands our ability to better assign functions to the newly predicted coding sequences. Comparative analysis of sequences for multiple genomes will provide substantially more information on the emerging and re-emerging new biological toxins and super-antigens, and this information will be very valuable in the discovery of new signature sequences to enhance bio-detection, protection and treatment. A model comparative analysis using the complete genome sequence of an M1 strain of Streptococcus pyogenes, also known as group A streptococci (GAS) which is a strict human pathogen with no other known reservoir pr affected species will be discussed.

1. Introduction

Streptococcus pyogenes is a human pathogen, which is responsible for a wide variety of diseases, including pharyngitis (streptococcal sore throat), scarlet fever, impetigo and toxic shock syndrome [1,2]. Genetic variability is known to occur, as evidenced by the appearance of strains associated with outbreaks of infection such as necrotizing fasciitis, toxic shock syndrome and rheumatic fever [1–3]. Most of the human and animal biological threat is shown in schematics 1. Most recently in 2002 there was an outbreak of severe acute respiratory syndrome (SARS) which started in china and spread all across the world. Three biological threats from the biological toxins such as botulinum toxin, staphylococcal enterotoxin B and ricin are also listed in the schematics.

Human diseases	**Animal diseases**
Smallpox	African Swine Fever
Cholera	Foot and Mouth
Shigellosis	Fowl Plague
	Newcastle
	Rinderpest

* Present address: Dr. Akbar S. Khan, DTRA-CBM, 8725 John J. Kingman Road, MS 6201, Fort Belvoir, VA 22060-6201; Akbar.Khan@DTRA.MIL.

Zoonoses

Melioidosis	**Anthrax**
Glanders	**Brucellosis**
Plague	**Coccidioidomycosis**
Psittacosis	**EEE / VEE / WEE**
Q Fever	**Japanese B**
Rabies	**Ebola/Marburg**
Tularemia	**Histoplasmosis**
	SARS

Botulism
SEB intoxication
Ricin

Figure 1. Schematic representation of all the biological threats and their inter-relationship among each other.

Table 1. Level of toxicities or infectivityies of different biological agents and their comparison to toxicity of nerve agent VX.

	Respiratory Dose for Man (ug)	
Nerve Agent VX	8,000	↓ -More infective -More Toxic
Botulinum A	4.5	
SEB	0.025	
Anthrax	0.008	
VEE	0.0004	
Tularemia	0.0001	
Q Fever	0.0000021	

Comparative list of toxicity of biological threat agents:

A comparative list of biological threat agents is shown in Table 1. This table compares the toxicity of biological threat agents with Nerve Agent VX to point extreme toxic and infective nature of these biological threat agents.

New emerging toxin and super-antigen like genes in Streptococcus pyogenes Genome

Putative virulence associated genes are abundant in the genome of Streptococcus pyogenes M1 strain, many of the encoded proteins predicted to be localized to the cell surface or secreted as extracellular products. At least six genes encoding new superantigen-like proteins also are found, many of which are associated with mobile genetic elements, making a total of 14 superantigen-like molecules identified to date in this genome.

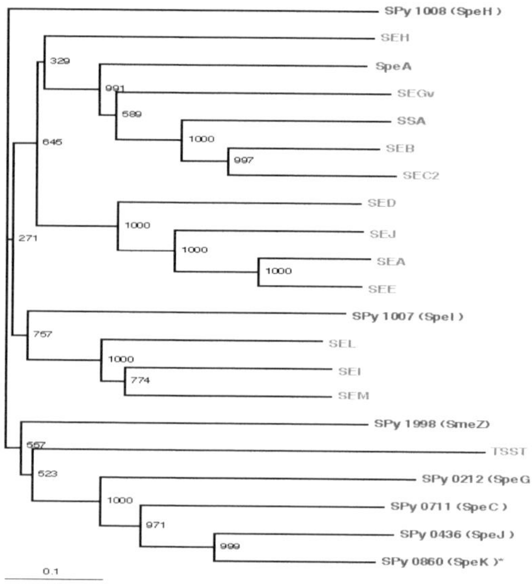

Figure 2. Phylogram showing the relationship among superantigen like-proteins in *S. pyogenes* SF 370 genome. The figure is reproduced from PNAS (2001), 98, 4658–4663.

```
>ref|NP_664724.1| streptococcal superantigen SSA - phage associated
[Streptococcus pyogenes phage 315.2]

 ref|NP_002381.1| SSA precursor [Streptococcus pyogenes SSI-1]
         Length = 260

 Score = 307 bits (787), Expect = 1e-84
 Identities = 155/263 (58%), Positives = 198/263 (75%), Gaps = 10/263 (3%)

Query: 4    RLFISHVILIFALILVISTPNVLAESQPDPKPDELHKSSKFTGLMENMKVLYDDNHVSAI 63
            R+ +   ++  A +L IS   V A SQPDP P++L+KSS+FTG+M N++ LYD++ V
Sbjct: 6    RILVVACVVFCAQLLSIS---VFASSQPDPTPEQLNKSSQFTGVMGNLRCLYDNHFVEGT 62

Query: 64   NVKSIDQFLYFDLIYSIKDTKLGNYDNVRVEFKNKDLADKYKDKYVDVFGANYYYQCYFS 123
            NV+S  Q L  DLI+ IKD KL NYD+V+ EF +KDLA KYK+K VD+FG+NYYY CY+S
Sbjct: 63   NVRSTGQLLQHDLIFPIKDLKLKNYDSVKTEFNSKDLAAKYKNKDVDIFGSNYYYNCYYS 122

Query: 124  KKTNDINSHQTDKRKTCMYGGVTEHNGNQLD-KYRSITVRVFEDGKNLLSFDVQTNKKKV 182
            + +   N+     +KTCMYGGVTEH+ NQ++ K+ +ITV+V+ED +N+LSFD+ TNKK+V
Sbjct: 123  EGNSCKNA-----KKTCMYGGVTEHHRNQIEGKFPNITVKVYEDNENILSFDITTNKKQV 177

Query: 183  TAQELDYLTRHYLVKNKKLYEFNNSPYETGYIKFIENE-NSFWYDMMPAPGDKFDQSKYL 241
            T QELD  TR  LV  K LYEFNNSPYETGYIKFIE+  +SFWYDMMPAPG  FDQSKYL
Sbjct: 178  TVQELDCKTRKILVSRKNLYEFNNSPYETGYIKFIESSGDSFWYDMMPAPGAIFDQSKYL 237

Query: 242  MMYNDNKMVDSKDVKIEVYLTTK 264
            M+YNDNK V S  + IEV+LT K
Sbjct: 238  MLYNDNKTVSSSAIAIEVHLTKK 260
```

Figure 3. Protein-protein blast of *Streptococcal* superantigen (SSA) with *Streptococcal enterotoxin B* protein sequences.

These known or putative streptococcal proteins all have at least one related protein identified from another Gram-positive bacterial species, suggesting that these genes may have been originated by horizontal transfer as shown in Phylogram in Fig. 3.

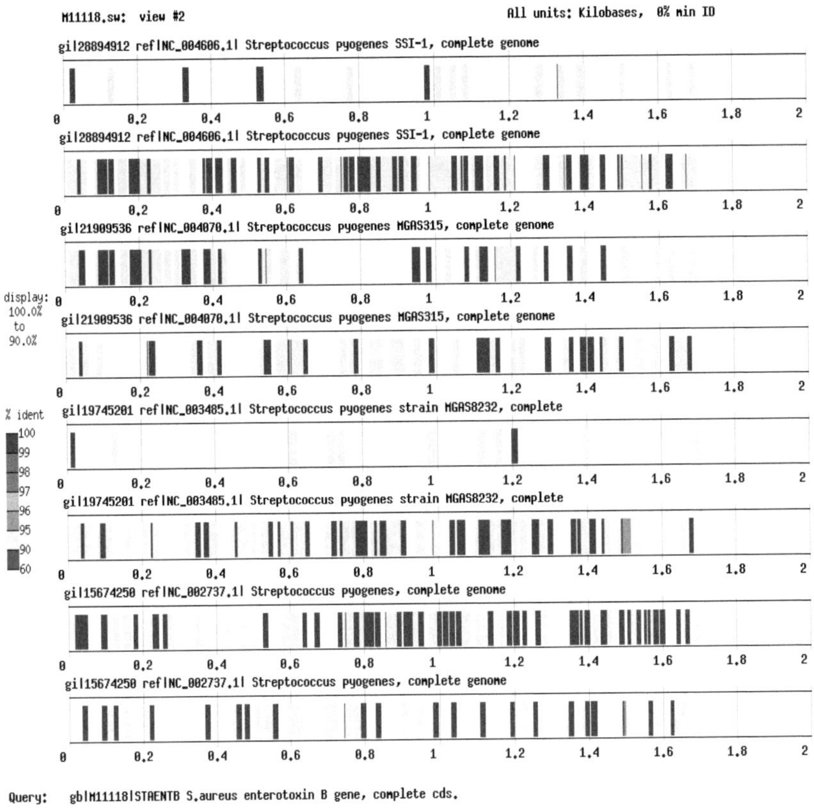

Figure 4. Comparative analysis of multiple strains of *Streptococcus pyogenes* with *Streptococcal enterotoxin B* gene with cross-match programs from Southwest Parallel Software, Inc.

A comparative protein-protein blast analysis of *Streptococcal* super-antigen (SSA) with *Streptococcal* enterotoxin B protein sequences as shown in Fig. 3 showed 50% identities in the sequences strongly implicating the presence of SEB toxin protein like sequences in the genomes of *Streptococcus Pyogenes*.

Futhermore, a cross-comparison of genomes of different strains of *Streptococcus pyogenes* with SEB gene showed a high level of identity and similarity as shown in Fig. 4. These comparative analyses further confirmed the fact that superantigen and protein toxin like sequences or genes exits in the newly sequenced genomes of *Streptococcus pyogenes*.

Conclusions

The threat of biological warfare and terrorism is a chilling reality and the genomic and proteomic revolution has the potential to have major impacts on this threat during this 21st century. It is the major responsibility of global biomedical community to develop and fund programs that will address the threat of biological weapons, its prevention, detection and deterrence.

References

[1] Ferretti J.J., McShan W.M., Ajdic D., Savic D.J., Savic D., Lyon K., Primeaux C., Sezate S., Suvorov A.N., Kenton S., Lai H.S., Lin S.P., Qian Y., Jia H.G., Najar F., Ren Q., Zhu H., Song L., White J., Yuan X., White S., Roe B.A. and McLaughlin R. Complete genome sequence of an M1 strain of *Streptococcus pyogenes*. PNAS 2001; 98, 4658–4663.

[2] Nair G.B. and Takeda Y. The heat stable enterotoxins. Microb. Pathog; 1998; 24, 123–131.

Emerging Biological Threat
G. Berencsi et al. (Eds.)
IOS Press, 2005

Application of Genomics and Proteomics for Detection Assay Development for Biological Agents of Mass Destruction

Vito G. DELVECCHIO
Institute of Molecular Biology and Medicine, University of Scranton, Scranton, Pa. 18510, Vital Probes, Inc. 1300 Old Plank Road, Mayfield Pa, 18433

Distinguishing a particular microorganism from all others demands finding a target sequence that is unique for that organism, is conserved in all members of that species or strain, and has limited or no plasticity. Such targets can be a nucleic acid sequence, constitutive protein, or an antigenic epitope. The target can be probed with PCR-based, immunochemical, or mass spectroscopy assays. Genomics has played an important role in developing new generations of probes. The genomes of an organisms of which is the nucleic acid content has been sequenced and annotated is of invaluable aid in the identifications of targets. Post genomic disciplines such as proteomics and glycomics are and will usher in a new generation probes. The proteome can be defined as a set of proteins produced by an organism under a defined set of conditions. The presence of a protein is the ultimate proof that a gene is being expressed. The glycome represents the glycan groups or saccharide chains attached to proteins or lipids. Surface proteins (S-layer) are ideal targets. Since they are often found at the surface of a cell they are probed with antibodies without breaking the cell.

Genomic strategies that have been used in probe development assays include the uses of bioinformatics and *in silico* studies, fingerprints analysis, and suppressive subtractive hybridization (SSH). Proteomic approaches that are now emerging include global analysis in which the entire protein compliment is studied, comparative proteomics whereby the proteome of one species is compared to a closely-related species. Comparative proteomics has been used to investigate the secretome, which is the set of proteins secreted into a defined or semi synthetic medium. The secretome is sometimes referred to as the culture filtrate. The proteomes of an organism grown under different conditions can also yield valuable data for probe development. The protein content of a specific part of a cell such as the exosporium or S-layer is of great interest for these are often involved in attachment, internalization, and harming of host cells.

SSH is a PCR-base technique in which the genomic DNA of the organism of interest, referred to as the tester DNA, is hybridized with a reference DNA, termed the driver. The hybridized sequences that are common to the tester and the driver are then removed. The unhybridized sequences are tester-specific. These tester sequences are then cloned into *E. coli* for further investigation. SSH has been used to identify *Bacillus anthracis* chromosomal sequences that are not present in the closely-related *B. cereus* and *B. thuringiensis*. Figure 1 shows the relative locations of the *B. anthracis*-specific open reading frames (ORFs).

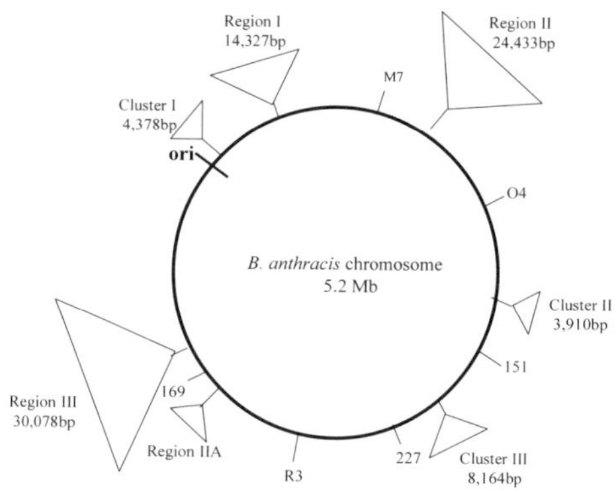

Figure 1. *B. anthracis*-specific chromosomal components.

Various sequence found in these regions, cluster or isolated orphan ORFs have been observed by PCR amplification to be ideal targets for the identification of *B. anthracis.* Region I is composed of ORFs that had similarity with genes encoding enzymes and proteins involved in cell wall surface polysaccharides. Regions II and III contained bacteriophage-related sequences including integrases and recombinases.

The presence of a gene or ORF does not necessarily mean that a protein is produced. In fact only a small percentage of tRNAs are translated into proteins. Thus the presence of a protein is the ultimate proof that a gene or ORF is capable of being expressed. Proteomics has been the mains strategy for investigating gene expression on a global or comparative level. It is the premier post-genomic discipline.

Standard proteomics studies utilize two-dimensional gel electrophoresis (2GE) in which the proteins are separated by isoelectric focusing (IEF) in the first dimension and by SDS-PAGE in the second dimension. 2GE is a powerful technique for resolving complex mixtures of proteins and permitting the simultaneous analysis of hundreds or thousands of proteins. IEF separates proteins on the basis of their isoelectric point (pI). The pI of a protein is the pH at which the protein has a zero net charge. When a protein mixture is applied to an IEF gel, each protein migrates until it reaches the pH that matches its pI. Proteins with different pIs align at different positions throughout the gel. In the conventional method, proteins are separated in a pH gradient generated by applying an electric field to a gel containing a mixture of free carrier ampholytes. Carrier ampholytes are low molecular mass components (i.e., organic acids and bases) with both amino and carboxyl groups. When an electric field is applied to a solution of ampholytes, positively charged molecules migrate toward the cathode and those that are negatively charged move toward the anode generating a defined pH gradient. At its isoelectric point, an ampholyte cannot migrate because it has a zero net charge. When an IEF gel with a stable pH gradient is used, separation is achieved when each protein molecule migrates to the position of its isoelectric point and accumulates (focuses) there. Since the pH at each end of the gel is known, IEF is used to determine the isoelectric point of a particular protein.

After IEF, proteins are separated in the second dimension by electrophoresis in the presence of the detergent sodium dodecal sulfate (SDS). SDS binds to most proteins by hydrophobic interactions in amounts proportional to the molecular mass of the protein. The bound SDS contributes a large negative charge, rendering the intrinsic charge of the protein insignificant. Furthermore, when SDS is bound, most proteins assume a similar shape, resulting in a similar ratio of charge to mass. Thus, protein separation is based almost exclusively on the basis of mass, with smaller polypeptides migrating most rapidly.

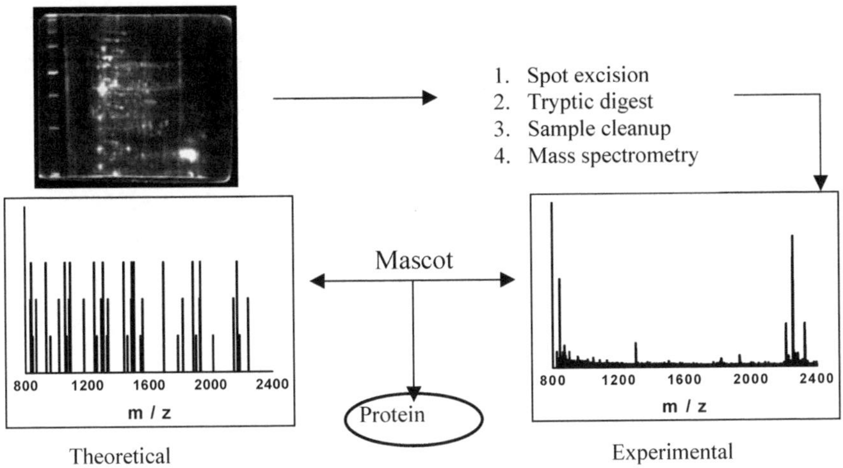

Figure 2. Method of protein identification using the Mascot search engine.

After 2GE, the proteins are visualized by staining with SYPRO® Ruby, a ruthenium-based fluorescent stain shown to have several advantages over other commonly used protein stains with respect to sensitivity and linear dynamic range. The stained gels are imaged and the spots are picked for protein identification. The protein spots are subjected to protease (e.g. trypsin) digestion to yield smaller peptide fragments, the number and size of which are characteristic for a particular protein. The digested protein is then spotted onto a plate and identified by mass spectrometry (MS).

One commonly used instrument for most MS work is Matrix-Assisted Laser Desorption Ionization Time of Flight (MALDI-TOF) Mass Spectrometer. The mass analyzer resolves ions based on their mass/charge (m/z) ratio that is proportional to their velocity. This is dependent upon the time required for the ions to travel the length of the flight tube, i.e., the time between application of voltage to the plate and the registration of signal by the detector. The smaller the m/z value, the shorter is the flight time and the faster the ions reach the detector. The detector then converts the kinetic energy of the arriving particles into electrical signals. Resolution in mass spectrometry refers to the ability of the instrument to distinguish between ions of slightly different m/z values.

The mass spectrum generated from MALDI-TOF-MS is called a mass fingerprint which is characteristic for a particular protein. The identity of an unknown protein can be determined by comparing its peptide mass fingerprint with the theoretical spectrum generated by digestion of each of the proteins in a database by using a search engine such as Mascot. Protein identification from enzymatically-derived peptides depends on the frequency of specific cleavage sites within a protein. The cleavage sites yield a set of potential peptide masses that are unique to that sequence entry when compared to all other proteins entered in the database. A protein is identified when a significant number of the experimentally determined peptide molecular weights match the m/z values in the theoretical mass spectrum. Some of the databases commonly used for protein identification include NCBInr, SWISS-PROT, TrEMBL and OWL. Thus, a rapid and automated protein characterization is achieved. Alternatively and more advantageous in terms of speed and accuracy is matching the MS spectra with translations of the nucleotide sequence from an annotated genome of the same organism. This enables investigator to match expressed proteins to their corresponding open reading frames (ORFs) (Fig. 2). For instance, the availability of a completely sequenced and annotated *B. melitensis* genome has paved the way for a highly comprehensive and rapid analysis of its proteome [1].

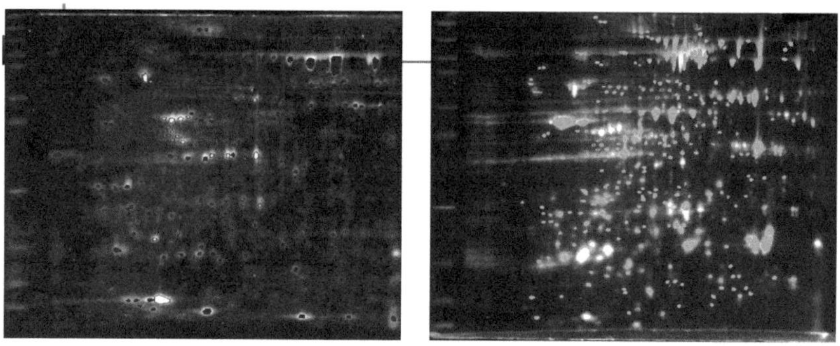

Non-induced (254 spots) **Induced (322 spots)**

Figure 3. Overview of *Bacillus anthracis* RA3 secretomes (pH 4–7).

The results of proteome analysis indicate which genes are expressed under a given set of conditions, how protein products are modified, and how they might interact. Unlike the genome, the proteome is not static. It changes with the state of development, under conditions of environmental stress, and during disease states of a tissue. There are many more proteins in a proteome (mainly due to posttranslational modifications) than genes in a genome.

It is now possible to bridge the gap between genotype and phenotype. Using proteomics, we are not only able to determine what gene or ORF is responsible for a particular virulence property of a pathogen but also whether or not its product or protein is being expressed. Signature proteins can be applied to immunochemical assays to determine if a particular gene is being expressed. The gene that codes for that particular protein can be identified by various nucleic acid-based amplification assays. One must constantly keep in mind that the presence of a gene or its mRNA does not translate into the expression of its protein. Thus both nucleic acid- and protein-based assay must be used to determine if a particular virulence factor is present in a particular biological weapon of mass destruction (BWMD). This is extremely important in determining if a foreign virulence gene has been engineered into a BWMD.

Knowledge of proteins that are induced during infection and those that contribute to pathogenicity would aid in the design of safe, efficient vaccines against *B. anthracis*, and could lead to the discovery of a new generation of effective anti-anthrax drugs. Such biomarkers could also be of immense aid in specific probe assays for *B. anthracis*. Using current proteomics technology, the proteins secreted into the growth medium in large quantities defined as "secretome" of the virulent *B. anthracis* have been analyzed under conditions that simulate an *in vivo* host cell. These conditions include the use of R-medium [2] that has been demonstrated to induce anthrax toxins and are totally synthetic so it facilitates the direct isolation of extracellular *B. anthracis* proteins. Secreted proteins have been isolated from the synthetic medium [3,4]. Preliminary results indicate that twenty-seven proteins were detected in the simulated secretome. The well-characterized toxins, lethal factor, edema factor, the protective antigens as well as the known surface proteins were identified. In addition, a chromosomal *B. anthracis* enolase was identified which corresponded to an immunoreactive protein in *Streptococcus mutants*. Overall, these findings could lead to the development of a new immunodetection assay for *B. anthracis* and in the identification of stress factors and virulence-associated secreted proteins [5]. However, more investigations are warranted. Figure 3 shows the proteomes generated in induced and uninduced conditions.

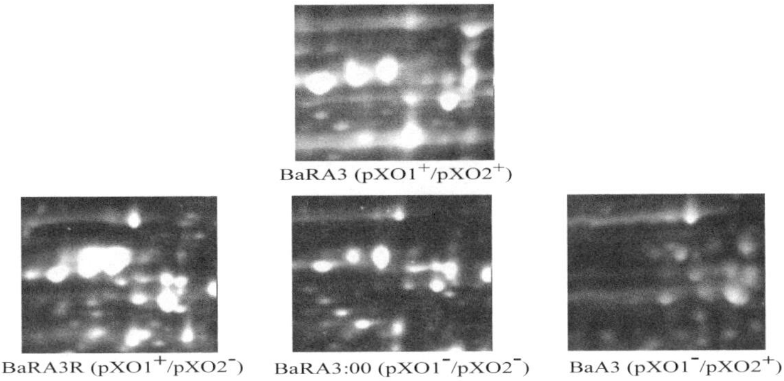

Figure 4. Induced Secretomes of Virulent and Avirulent *Bacillus anthracis*.

The secretomes of fully virulent, strains containing pXO1 but cured of pXO2; strains with pXO2 but no pXO1, and fully cured with no plasmids has been investigated to understand the pathogen response to simulated host conditions. Figure 4 is a representation of part of the secretome of the various cured and uncured strains.

Brucellosis is a major zoonotic disease that causes abortion and sterility in wild and domesticated animals and Malta fever in humans. Although the spread of the disease is controlled in developed countries by livestock testing, vaccination, and slaughter programs, brucellosis continues to be a major problem in the Mediterranean region and parts of Asia, Africa and Latin America where it causes severe economic losses [6].

B. melitensis, *B. abortus*, *B. suis* are the most common causal agents of brucellosis in humans. The World Health Organization considers *B. melitensis* as the most important zoonotic agent. It is extremely infectious (1–10 cfu per person) partly due to its highly aerosolic nature [7]. Because of its ability to cause a debilitating disease in humans, it is considered a potent biological warfare agent. The *Brucella* genus is composed of six currently recognized species: *B. melitensis*, *B. abortus*, *B. suis*, *B. ovis*, *B. canis*, and *B. neotomae* [6,8]. Each species show preference in their host specificity to other animals. Recently brucellosis has been described in a variety of marine mammal throughout the world. Brucella are gram-negative, nonmotile, non-spore forming, coccobacilli [6]. The genome of one of its species, *B. melitensis* strain 16M, has recently been sequenced, closed, and annotated [1]. The genome is composed of 3,294,931 bp distributed over two circular chromosomes of 2,117,144 bp and 1,177,787 bp and predicted to encode for 3198 open reading frames (ORFs).

The global proteome of *B. melitensis* grown under laboratory conditions has been examined and is still a work in progress [1,4,9]. Figure 5 is a representation of the ORF expressed in laboratory grown cultures of *B. melitensis*.

Secreted proteins often play an important role in the pathogenicity of microorganisms. It is therefore advantageous to identify which proteins are exported by a pathogenic bacterium and to acquire knowledge concerning how those proteins are exported from the cell. By genomic analysis, *Brucella* species are known to contain genes which share homology to those of known type I, III, IV, and V secretion systems. Brucella contains a conserved operon, *virB*, which has homology to type IV secretion system genes. In addition, 400–600 proteins of the *B. melitensis*, *B. suis*, and *B. abortus* theoretical proteomes are predicted to contain signal sequences which direct them to the periplasmic space, the outer membrane, or for export from the cell. While genetic analysis and computer modeling can provide insight to the potential of an organism, empirical evidence is necessary to confirm *in silico* hypothesis.

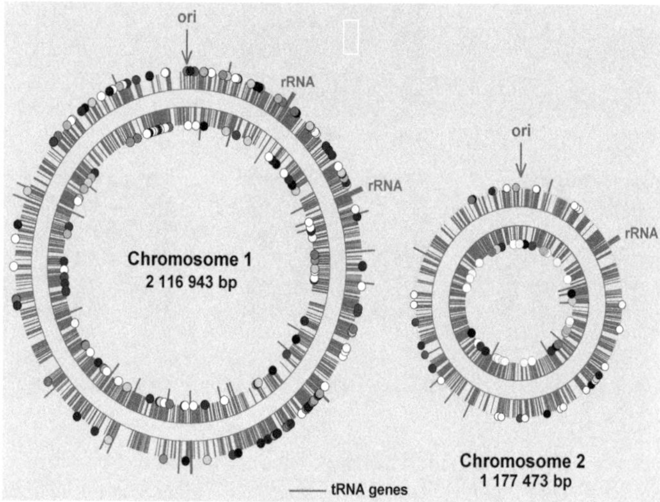

Figure 5. Linear representation of what proteins are expressed under laboratory conditions by *B. melitensis.*

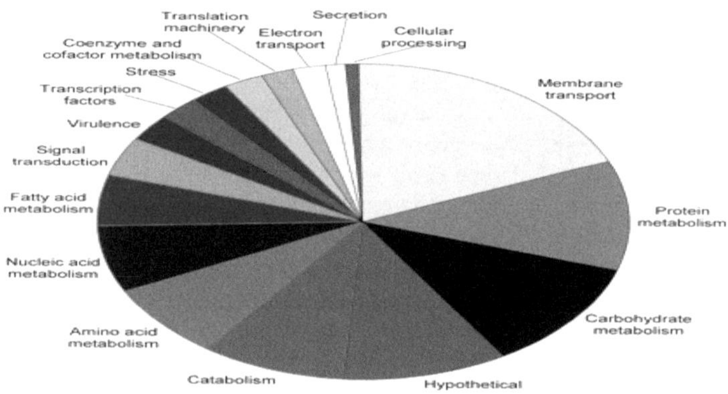

Figure 6. is an illustration showing the various classes of proteins represented by the different colored spheres.

Such empirical evidence can be obtained by proteomic studies of the Brucella secretomes. Preliminary investigations have indicated that using a semi synthetic medium, which contains a minimum of complex biological nutrients, such studies can be achieved. Figure 7 is a 2DG of the secretome of various mutants of the virulence operon of *B. abortus*.

Membrane proteins are of special interest in the pathogenic process. The surface proteins are often the first to interact with the host cells. Many of the offensive mechanisms are found on the surface layer of a microorganism. Epitope biomarkers are easy to detect since the pathogen does not have to be lysed to probe. Surface proteins are often ideal vaccine targets. Figure 8 is a 2DG of total cellular proteins as compared to those found in the membrane of *B. melitensis*.

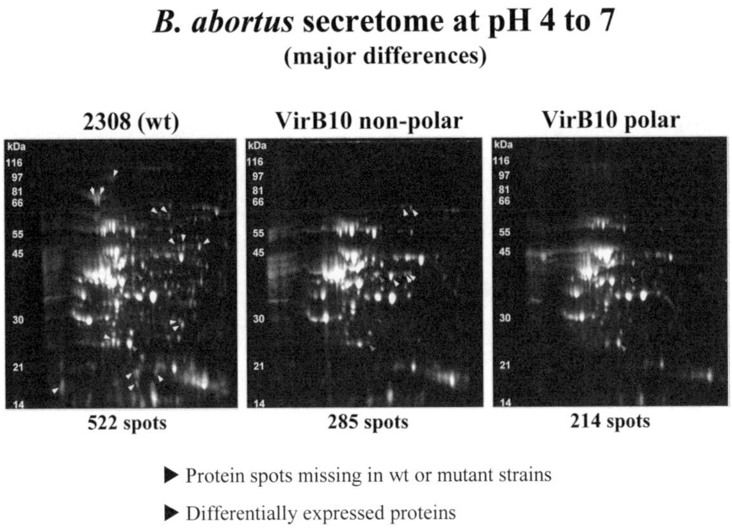

Figure 7. Secretome of wild type and virulence operon mutants.

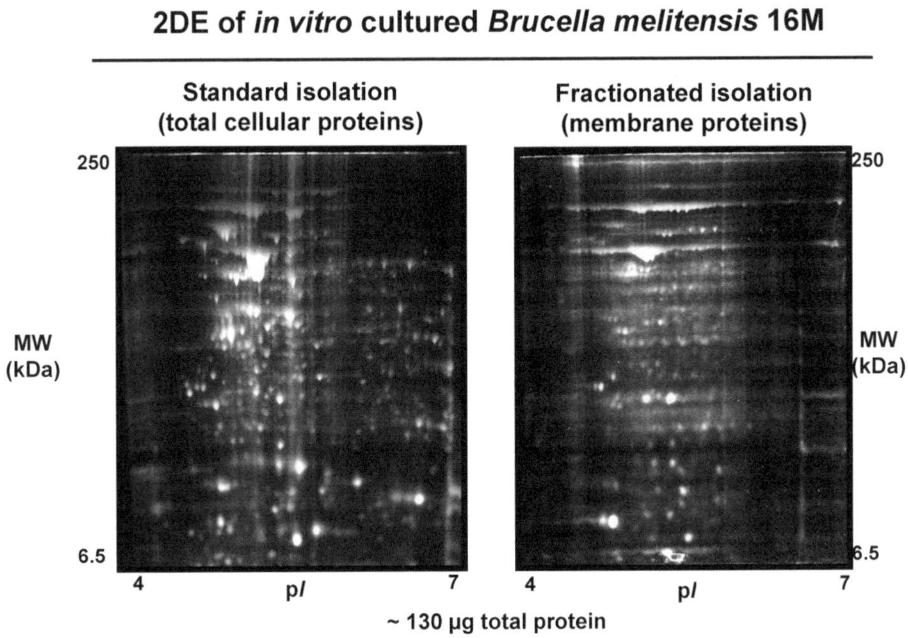

Figure 8. Standard and membrane proteins of *B. melitensis*.

Once proteomics investigations have been identified potential biomarkers, probes can be developed to target specific nucleic acid and protein. Figure 9 is an illustration of this strategy.

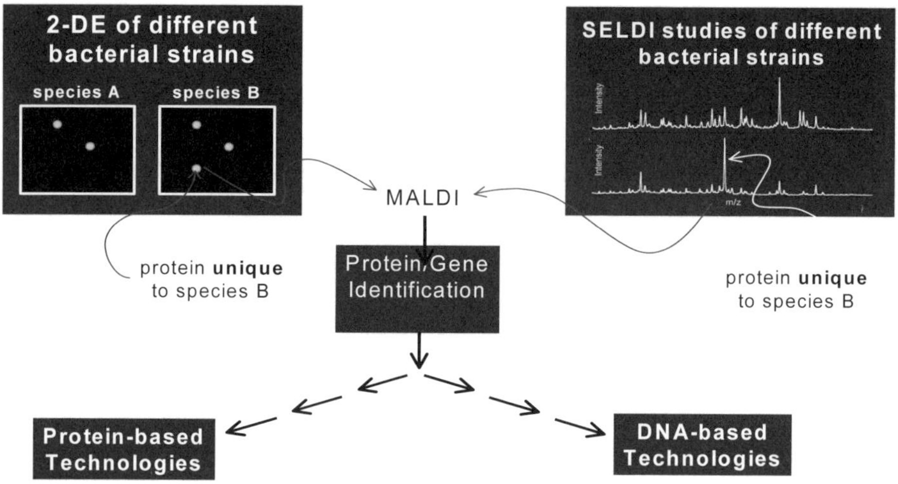

Figure 9. Proteomic biomarker discovery strategy.

References

[1] DelVecchio, V.G., V. Kapatral, R.J. Redkar, G. Patra, C. Mujer, T. Los, N. Ivanova, I. Anderson, A. Bhattacharyya, A. Lykidis, G. Reznik, L. Jablonski, N. Larsen, M. D'Souza, A. Bernal, M. Mazur, E. Goltsman, E. Selkov, P.H. Elzer, S. Hagius, D. O'Callaghan, J.J. Letesson, R. Haselkorn, N. Kyrpides, and R. Overbeek.(2002) The genome sequence of the facultative intracellular pathogen *Brucella melitensis*. Proc. Natl. Acad. Sci. USA. 99:443–448.

[2] Ristroph, J.D. and B.E. Ivins. (1983) Elaboration of *Bacillus anthracis* antigens in a new defined culture medium, Infect. Immun. 39:483–486.

[3] Radie-Kolpin, M., R.C. Essenberg, J.H. Wyckoff. (1996) Identification and comparison of macrophage-induced proteins and proteins induced under various stress conditions in *Brucella abortus*. Infect. Immun. 64:5274–5283.

[4] Wagner, M.A., M. Eschenbrenner, T.A. Horn, et al., (2002) Global Analysis of the *Brucella melitensis* proteome: identification of the proteins expressed in laboratory-grown culture, Proteomics 2:1047–1060.

[5] Patra, G., Williams, L.E., Dwyer, K.G. and DelVecchio, V.G. (2003) *Bacillus anthracis* secretome. *In* Applications of Genomics and Proteomics for Analysis of Bacterial Biological Warfare Agents. Eds V.G. DelVecchio and V. Kremery. NATO ASI Series.

[6] Corbel, M.J., W.J. Brinley-Morgan, in Bergey's Manual of Systematic Bacteriology, Eds. N.R. Krieg and J.G. Holt Williams and Wilkins (1984) 377–388.

[7] Miller, C.D., J.R. Songer, and J.F. Sullivan (1987): A twenty five year review of laboratory-acquired human infections at the National Animal Disease Center, Am. Ind. Hyg. Assoc. J. 48 271–275.

[8] Verger, J.M., F. Grimont, P.A. Grimont, and M. Grayon. (1987) Taxonomy of the genus *Brucella*. Ann Inst. Pasteur Microbiol. 138:235–238.

[9] Mujer, C.V., M.A. Wagner, M. Eschenbrenner, T.A. Horn, J.A. Kraycer, R. Redkar and V.G. DelVecchio. (2002) Global analysis of *Brucella melitensis* Proteomes: its potential use in vaccine development, identification of virulence proteins and establishing evolutionary relatedness. *In*: "The Domestic Animal/Wildlife Interface: Issues for Disease Control, Conservation, Sustainable Food Production, and Emerging Diseases." E. Paul J. Gibbs and Bob H. Bokma (Eds.). Ann. N.Y. Acad. Sci. 969:97–101.

Emerging Biological Threat
G. Berencsi et al. (Eds.)
IOS Press, 2005

Development of Reagent Kits for Detection of Lethal Toxins

Gábor FALUDI[a], István JANKOVICS[b], Ildikó VISONTAI[c], Júlia SARKADI[b] and Gyöngyi ZELENKA[a]
[a] *Institute of Health Protection, Hungarian Defence Forces*
[b] *OMNINVEST Ltd., Department for Quality Assurance*
[c] *B. Johan National Center For Epidemiology, Budapest, Hungary*

The "mysterious world of toxins" is connected not only to the force protection of military units, but also to our normal civil life. Several historical examples are important steps of our learning on microbial toxins. The illnesses caused by food poisoning during the last 5 years in Hungary are briefly mentioned [1,2]. Possible threat of emerging microorganisms has to be also taken into account. The demand for rapid detection has initiated the local development of reagents for the rapid detection of RICIN and *S. aureus* enterotoxin B. Emerging bioterrorism has to be coped by networks of military and civil laboratory facilities to be prone in preventing biological disasters.

Introduction

Curativ role of the toxin of *C. botulini* (Botox) in the therapy has revealed new aspects of microbial toxins than that in classical microbiology. The discovery and characterisation of new and new microbial and plant toxins resulted in the preparation of weaponized forms, too. These have to be considered as a great challange not only for military experts but also for preventive medical specialists. The new emerging threate "toxins in the hands of terrorists" have to be also taken into account and the threat made the authors aware to follow the NATO recommendations as far as the "Hand Held Test Kit" philosophy is concerned [3–6].

1. Toxins' Role in the Past

Each country and geographical region posesses specific events of its own tragic history [7–9]. "The mongolian invasion" has been one of them in the Carpathian Basin, which nearly destroyed the ancient Hungarian Kingdom and killed a large proportion of the people in the 13th century. This successfull invasion was the second campaign of Khan Batu son of Ghengis, who had to postpone the start of his great western offensive planned to begin a year before. During his first campaign the mongolian troops – which were called that time by us: Tatars – could not collect enough fresh food for their huge herds of horses in the Russian plains. The offence had to be postponed, since they were constrained to utilize the several years old thatches of roofs of villages, what was impregnated with large amount of accumulated mycotoxins. Consequently they have lost a considerable part of their horses. The first date and plan of invasion frustrated because of the lack of mobility [7–10].

Table 1. Food intoxications between 1997 to 2002 in Hungary [1,2].

Origin of toxins/date	1997	1998	1999	2000	2001	2002
C. botulini	100	92	142	9	130	170
B. cereus	28	30	31	10	22	5
S. aureus	26	76	265	126	295	209
Mycotoxins	261	555	145	171	300	252
Veterinary origin	0	4	0	0	0	0
Plant origin	0	0	0	4	2	2

2. Toxins in Our Everyday Life

The toxins play important role all around the world in the emerging threat of food intoxications, because of the large transnational chains of food supply and consumption. The global transportation systems of rice, coffe, oilic seeds, and spices render the mycotoxin exposition possible everywhere [8–10].

The national epidemiological features of food intoxications in Hungary are similar to the international trends. The food borne infections and intoxications has been notifiable since 1951 in Hungary. On the basis of collected data of the National Institute of Food Hygiene and Nutrition [1,2] the tendencies could be easily determined. The role of heat-stabile staphylococcal enterotoxins was dominant in the 60's and 70's. Considerable proportion (24–30%) of the infections have been transmitted by home made haedcheeseee from pigs, and the same bacterial strains could be detected and identified from the throat or hands of kitchen workers.

Nowadays about ten outbreaks occur yearly. The data from the last 6 years (1997–2002) are shown in Table 1.

The annual incidence of botulism – documented regularly from the 60's – were at a low level, 1 case per 3–4 years. In the 70's and 80's this number increased up to 7–8 events per year, and this number seems to be countinous. The number of patients registered during the last 6 years are shown in Table 1.

The vehicle was shown to be the haedcheeseee again, the territorial distribution, however, was extremely characteristic. Seventy per cent of the patients were concentrated in Baranya county near to the most southern border of Hungary (previous Yugoslavia at present Serbia-Montenegro).

A second important field of toxicology was water hygiene. The quality of water purification technology in combination with the quality of the nataral water sources resulted occasionally clinical illnesses. Professionals had to face several times with many uncomfortable skin symptoms among resting people at the Lake Balaton or at a few other waters.

Occasisonally due to the coincidence of several unique envionmental conditions (higher concentration of phosphates and so on) the blossom of algae were responsible for toxic symptoms. The boom was emerged under typical seasonal circumstances. The critical concentration of algal cells reached occasionally 10.000 cells/ml. Late summer the boom of blue-green algae and the presence of their toxic byproducts (i.e. circular polipeptides composed of d-amino acids) could be detected. Microcystin and anatoxin were also detected in the waters in certain years.

3. Toxins for Military Application

"Toxins are poisonous materials produced by different types of organisms" is a definition taken from the FM 3–9 document for toxic agent of biological weapons [3,9]. There were

Table 2. Classification of biological toxins by origin [6] and their chemical nature.

Natural origin	High specific toxicity	Main chemical character
Bacteria	29 of > 60	Proteins and enzymes
Fungi	26 of > 30	Organic substances, enzymes
Algae	2 of > 30	Circular peptides
Plants	5 of > 36	Heterogeneous
Arthropodes	2 of > 22	Enzymes and proteins
Marine organisms	65 of > 110	Heterogeneous
Amphibians	5 of > 5	Enzymes and proteins
Snakes	8 of > 124	Enzymes and proteins

lots of toxins discovered – the tentative number was about 400, but only a few of them was found to be suitable bioagents [9–11]. Very high specific toxicity, stability and the possibility of large-scale production were the main criteria of biowarfare ustilisation. The possible protection of corps was also a limiting standpoint previously, which has disappeared when the philisophy of bioterrorism emerged during the last decade.

These BW agents were also classified as bioterroristic agents by the CDC in 2000 according to the principles of risk assesment [11]. For many years this potencial of toxins seemed to be rather theoretical than practical. The reality of military use was too far from reality, while these weapons were kept under strict control by armed forces and international agreements. The general situation has gradually changed when the toxins got slide to the hands of terrorists. An illegal, uncontrollable and new form of proliferation was born as a dark side of our everydays. Unfortunately it is growing and spreading continously, evcen today. It has to be mentioned that in January 2003 6 possible Al-Qaida members were arrested by British authorities in UK for ricin production [9]. Similar event was published also from France recently by the media.

These two events of possible bioterrorism makes it obvious that toxins have to be included as an emerging threat together with other B and C agents. The detection, and verification of this toxin-threat has to be done also according to the NATO and CDC recommendations [6,7]. The final aim of the group was to prepare a "training ki" for the detection of toxins, which might help to educate the personnel in rapid diagnostic procedures.

4. Test for the Detection of Ricin and Staphylococcal Enterotoxin B

4.1. Material and Methods

The STANAG 4571 recommended colloidal gold system (Arista OD_{540}: 4) was introduced. Adsorbent pads AP-045 were purchased from Advanced Microdevices Ltd. UK, S70008, S70009 and S70010 from Pall, U.K. and 1Chr, 3MMChr, 31Chr, 17 Chr, CD427, D28, GF AVA or DVA and WF1.5 Whatman U.K., were used for the binding of antibodies.

The adsorption of staphyloccoccal enterotoxin B-specific antibodies was tested using Schleicher-Schüll FF85 and 125 filters, Millipore HF 07504, 13504, 18004 and 24004 filters.

Specific antibodies to ricin, rabbit IgG-specific goat antibodies and staphylococal enterotoxin B were purchased from SIGMA Hungaria.

One of the greatest problem was to find legal sources of antigens in order to prepare positive and negative controls and to perform optimalisation of the tests.

Under the international restrictive regulation of the VEREX Group the authors were compelled to prepare ricin from castor beans and SEB from liquid media of toxin producer

S. aureus. On the basis of these experiments it has to be declared, that the production of an efficient biological weapon is easier than its correct detection.

Modification of the legal background in 2000 in connection with the EU harmonization prevented the performance of certain working processes with BSL3 microorganisms. The conclusion of the authors was, that the legal situation of the terrorists is better, than that of the personnel working for the prevention of their activity.

4.2. Results

4.2.1. Ricin-Specific Detection Kit

Lateral flow diagnostic line-tests were developed for the detection of ricin and staphylococcal enterotoxin B. In addition to these a training kit was also developed, which makes possible the education of the laboratory personnel using innocuous samples and reagents.

The RICIN test is based on Immunopore -RP membranes of Whatman.

The test line (capture) contains RICIN-specific rabbit antibodies.

The control line is composed of rabbit-IgG-specific goat antibodies.

The conjugate is gold conjugated anti-ricin rabbit gamma globulin.

The sample volume was 100 microliter, and the detection limit was found to be 15 ng/ml ricin (i.e. 1.5 ng/100 microliter).

4.2.2. Detection Kit for Staphylococcal Enterotoxin B (SEB)

The SEB test is based on Immunopore -RP membranes of Whatman.

The test (capture) line contains SEB-specific rabbit antibodies.

The control line is composed of rabbit-IgG-specific goat antibodies.

The conjugate is gold conjugated anti-SEB rabbit gamma globulin.

The sample volume was 100 microliter, and the detection limit was found to be 15 ng/ml SEB (i.e. 1.5 ng/100 microliter).

4.2.3. Training Kit for Immunochromatographic Tests

The training kit will be patented in the near future. Therefore only the advantages can be presented here.

It is cheep.

No innocuous reagents are included and it is stimulating for the students of the training course.

4.3. Discussion

STANAG 4571 was successfully introduced in Hungary as far as ricin and SEB detection is concerned according to the recommendations of the “hand held reagent kits”. Detection of toxins of three other agents, however, anthrax, plague and botulism have to be developed, too.

Future tasks will be the validation, determination of the storage conditions and expiry dates of the kits.

The **training kit is an innovative progress** with innocuous reagent which might be a good tool in NBC training under conditions of normal life. The team could successfully solve the management of trading problems. Some of the harmful specific reagents, however, could be produced locally in small quantities in connection with the project.

Progress in the field can be performed by the replacemnet of polyclonal antibodies by the mixture of monoclonal ones.

Acknowledgements

We thank the help of Péter Major, Anna Szentmihályi, Géza Szita, Miklós Füzi and György Berencsi in providing publications of limited access, helpful comments, annual reports of institutions and comments during the preparation of the work. Technical assistance of Mrs. Eva Simon and Mrs. Magda Kaposi in the laboratory experiments is very much appreciated, too.

References

[1] Schiefner Kálmán, Törökné Kozma Andrea: Felszíni vizeink higiénés szemmel (Our surface waters through the eyes of a hygienist – in Hungarian). Egészségtudomány 41, 2 (1997) 145–151.

[2] Bejelentett élelmiszer-fertőzések, élelmiszer-mérgezések kórokozó szerinti megoszlása (1997–2001) (Microbial distribution of reported food poisonings in Hungary between 1997 and 2001). OKK Országos Élelmezés- és Táplálkozástudományi Intézet Évkönyve (Yearbook of the National Instituter of food and Alimentation). 2002.

[3] M. Dando, G.S. Pearson, T. Tóth: Verification of the Biological and Toxin Weapons Convention. NATO ASI Series. 2000. Kluwer Academic Publishers/Netherlands.

[4] E. Eitzen, J. Pavlin, T. Cieslak, G. Christopher, R. Culpepper: Medical management of biological casulties. 3 ed. 1998. Fort Detrick.

[5] R. Zajtchuk, R.F. Bellamy (Eds.): Medical Aspects of Chemical and Biological Warfare. Textbook of Military Medicine part I. 1997. USA.

[6] NATO Handbook on the Medical Aspects of NBC Defensive Operations (AmedP-6 (B) 1996. (STANAG 2500). Faludi G.: A biológiai fegyver jelentőségének megváltozása Modification of the role of biological weapons – in Hungarian). Honvédorvos. 1998. vol. 50. No. 1. 37–69.

[7] Dr. Faludi Gábor orvosezredes, Dr. Békési Lívia orvosőrnagy, Barabás Károly őrnagy, Prof. Dr. Halász László mérnökezredes, D.Sc.: A toxinok, mint biológiai harcanyagok (Toxins as biological warfare agents – in Hungarian). Honvédorvos. 1999. (51) 4. 192–210.

[8] Varga Ildikó, Matyasovszky Katalin, Sohár Judit: Élelmiszerek mikotoxin szennyezettségének jelentősége, adatok a hazai szintekről (Significance of the mycotoxin contamination of food at the Hungarian levels – in Hungarian). Egészségtudomány 44, 3 (2000) 224 – 241.

[9] Jeffrey M. Bale, Ph.D., Anjali Bhattacharjee, Eric Croddy, Richard Pilch, MD: Ricin Found in London: An al-Qa' ida Connection? CNS Reports.

[10] Kovács F.: Mykotoxinok a tápláléktláncban (Mycotoxins in the food chain in Hungarian). MTA (Hungarian Academy of Sciences). Budapest 1998.

[11] Faludi Gábor, Rókusz László: A biológiai fegyver (The biological weapon in Hungarian). Magyar Honvédség Egészségügyi Csoportfőnökség kiadványa (Publication of the Health Directorate of the Hungarian Defence Forces). 2003.

PART 3

BIOTERRORISM AS AN EMERGING AND REEMERGING BIOLOGICAL THREAT

Emerging Biological Threat
G. Berencsi et al. (Eds.)
IOS Press, 2005

Strategies for the Detection of Unknown Biological Materials

Peter J. STOPA[a] and Jeff MORGAN[b]
[a]*US Army Edgewood Chemical Biological Center, 5183 Blackhawk Road AMSRD-ECB-ENP-MC, E3549, Aberdeen Proving Ground, MD 21010-5424 USA*
[b]*Applied Ordnance Technology, Inc., 25 Center Street, Stafford, VA 22554*

One has to be first become aware that the possibility does indeed exist that "unconventional" biological agents be encountered in the field. There are a variety of strategies that could be implemented in either trigger or detection platforms that could be used to detect signatures from biological agents and possibly determine that they could present a danger to health and life.

However, we need to change our paradigms on how we think about the problem. The most important thing is to provide a detection and warning system that exploits credible signatures so that those in peril can take the appropriate protective measures.

1. Introduction

One of the keys to deterring the use of biological weapons is real time detection and warning. In the event that these agents are utilised, it is even more important to classify and eventually identify the type of agent so that appropriate countermeasures can be initiated. Several systems have been developed to provide detection and alarm of a biological warfare (BW) agent attack. These first generation systems detect the characteristics of an aerosol in order to measure changes in the aerosol content against a background. This may be indicative of a man-made (not naturally-occurring) event that could indicate a possible attack with a biological agent. These first generation systems then use antibodies to provide a means of characterization of the aerosol for specific types of biological materials. This approach presupposes a knowledge of what an adversary may have in their arsenal. With the advent of genetic engineering and thus the potential to design agents that may defeat current detection and identification strategies, additional or alternative signatures need to be exploited to reliably indicate a possible biological attack. This paper explores some additional signatures that can be used. Some approaches to improving medical intervention are also discussed.

1.1. Unknown Agents

What is meant by unknown agents? They are agents that are outside the sensor's library of known agents. They could be different agents or ones that have been genetically modified. This is even of more significance now that so many genetic sequences have been defined and so much information is available on the Internet. It is conceivable that in the not so distance future one could pick specific sequences to develop biological agents with defined

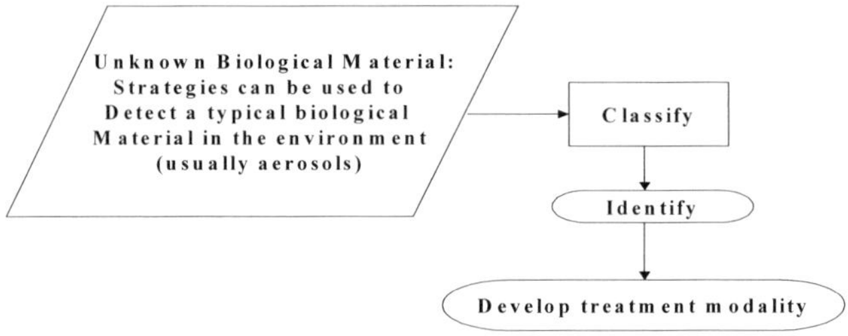

Figure 1. A strategy for response to a biological agent attack.

properties. Unknowns could also include encapsulated agents and/or next generation BW agents, ones that are not readily recognizable with today's reagents.

1.2. Rationale

The rationale for the importance of being able to detect and identify these agents is as follows. First of all, we need to differentiate between natural and non-natural (i.e., man-made events). A second reason is that we need to provide warning to troops and the civilian population that an attack may have occurred so that the potentially affected people can implement and maintain a protective posture. Likewise, one also needs to know when to retire from this posture. In terms of consequence management, one needs to be able to initiate the appropriate medical treatments for the affected population. And lastly, one needs to know the areas that need to be decontaminated and how effective that decontamination is.

2. Strategies for Detection

2.1. Trigger and Detector Startegies

A strategy for a response to a biological attack is illustrated in Fig. 1. It certainly would be beneficial to recognize a biological weapon when it is still in aerosol form. This would be important to "Detect to Warn" and also to trigger the sensor to collect a sample for further analysis. Logically, the next step would be to classify it as a bacteria, virus, protein toxin, or non-protein toxin. As a final step, you would want to identify it, which is important from a "Detect to Treat" perspective. Identification may also be necessary to prevent the outbreak of a disease or control secondary infections, especially if it is potentially communicable, such as plague, smallpox, Marburg, or Ebola virus. Identifying the agent could also be important for determining the proper medical response. Rapid tests could be performed to determine susceptibilities to various treatment regimes.

Currently particle size is the most widely used parameter that triggers an alarm, although technologies that measure particle shape, fluorescence from biological markers (tryptophan or NAD/NADH), or ATP luminescence are being developed and implemented into detection systems.

There are additional signatures that can be exploited. For example, elemental analysis can be performed to evaluate if a change in the ratio of various elements has occurred. Organic signature analysis can be performed to determine if materials that are consistent with various propellants, encapsulants, and aerosol additives are present. Changes in either the inorganic (elemental) or organic signatures could be indicative that a biological agent or

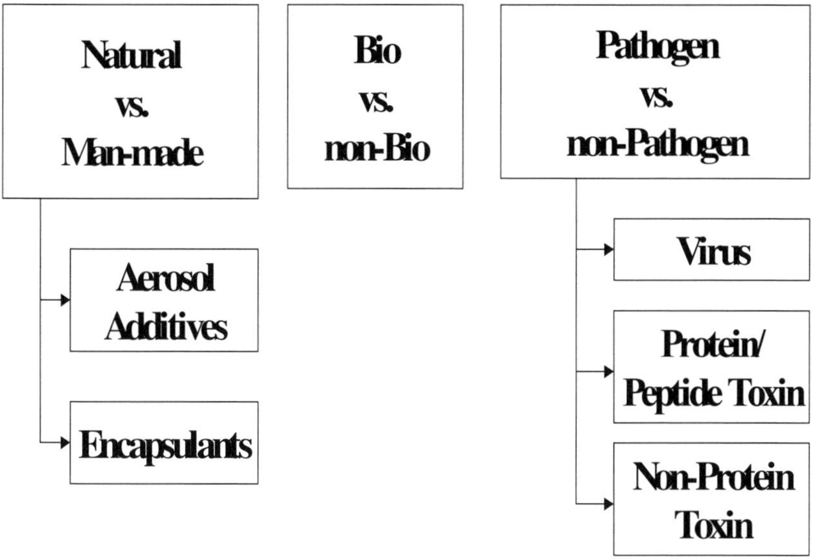

Figure 2. Classification parameters.

agents may have been used. These parameters can be used as an initial approach to determine if the background aerosol characteristics have changed; however, further characterization is warranted. A summary of these approaches is illustrated in Fig. 2.

As can be seen by the figure, there are a variety of strategies that one could employ in trying to classify the components of an unknown biological agent. First, there are several parameters that may be exploited to determine if an agent is either natural or man-made. These are shown in Fig. 3. Particle size is currently used by many fielded systems. Likewise, particle shape, especially the frequency of occurrence of a particular shape, could serve as a trigger. Elemental analysis is currently being considered by several groups as a means to determine whether a biological attack has occurred. For example, by determining the ratio of 2 or more elements, such as sodium, potassium, cobalt, zinc, etc., one might be able to determine if there has indeed been an attack. Calcium could be an indicator of spores, and nickel could be an indicator of botulinum toxin. There could also be additives in the mix, such as silica, diatomaceous earth, or cellulose, which ordinarily are not found in the atmosphere. The propellants that are used in a biological dissemination device may also have measurable signatures. An adversary may choose to encapsulate their weapon for a number of reasons. Encapsulants could be used to provide protection, to make the weapon more effective, or perhaps to cause false readings in sensors. Some technological approaches for the implementation of these strategies are shown Figs 3, 4, and 5.

2.2. Changes in Biological Flux

For example, it might be useful to see if there is a dramatic change in the biological flux of the environment. This is currently performed on existing platforms through the use of DNA measurements by flow cytometry and changes in the flux of Adenosine-Tri-Phosphate (ATP) by luminescence techniques. However, there are additional parameters that can be measured. For example, simultaneous DNA and Protein measurements have been shown to be effective to measure changes in the biological flux. Other parameters, such as heme

Parameter	Rationale	Test	Status
Particles	Most man-made aerosols are > 3 µm	Particle-size Analysis	Available (Field)
	May see uniform distribution of similar particle shapes	Particle-shape Analysis	Available (Field)
Elemental Analysis	May see "unnatural" elements or ratios of elements.	Spectrometric analysis	Available (Field) – Photoionization detectors (PO_4, SO_4) In Development – Sondebido Systems (Metals)
	May see spectra from silica, cellulose, or other organics	Active/Passive Standoff	Requires investigation
Propellants	May see propellants or other gases	Standoff; Gas Analyzers; IR FLIR; FTIR; Spectrometry	Requires investigation

Figure 3. Approaches for the determination of a suspect biological attack may contain a natural or man- made biological agent.

Parameter	Rationale	Test	Status
Silica particles	Silica particles to improve aerosol dissemination	Silica analysis	Available (Field)- Commertial manufacturers
Cellulose particles	Could be added as stabilizer/filler	Mass Spectrometry; Py-GC-IMS; Standoff	Under development
Protein contetnt	Protein content in air is low. Could be used as an additive or agent itself.	Standard Protein Determinations	Available (Field) – Various protein detection kits are routinely used.
	Could detect species markers from either additives or cells.	Histocompatibility Antigens	Available (Lab) – Used in FCM; could be adapted to other applications.
	Detect residual heme content from culture media, cells, etc.	Luminol Luminescence	Concept used during US BDWS program.
	Would have signatures from residual culture additives (e.g., vitamins and growth factors).	Mass Spectrometry	Available (Field)

Figure 4. Methods to detect aerosol additives or contaminants.

Parameter	Rationale	Test	Status
Particle Measurements	Encapsulated particles may have larger sizes, changes in refractive index, etc. that could be detected by scatter.	Refractive Index Shape Density Particle Size	Technology exists.
Absorbance	Encapsulants may impact the absorption spectra of biological materials.	Absorbance Spectra (UV/Vis/Ir)	Technology exists.
Composition	Some encapsulant materials may have unique signatures.	Mass Spectrometry Py-GC-IMS Elemental analysis	Technology exists.

Figure 5. Detection of encapsulant materials in biological agents.

measurements, viability changes, and a version of the Gram stain may also prove to be useful. Examples of this approach are seen in Fig. 6.

Parameter	Rationale	Test	Status
DNA / Protein Determination	Unknown agents will probably be present in particles that have measurable DNA / protein.	Flow Cytometer	Available (Field)
	Same as above; however measure total flux rather than only particulates.	Fluorometer	Available (Field)
Heme Determination	Biological materials will have heme present	Luminol Luminescence	Available (Field) – US BDWS Program.
Viability Determination	Viable biological materials have ATP present.	ATP Luminescence	Available (Field) – System Demonstrated
	Measure presence of lipases.	Fluoroscein Diacetate	Available (Field) – Routine reagent used in FCM and fluorescence.
Image Analysis	Perform Gram Stain with automated morphological analysis.	Conventional Microscopy	Gram Stain is widely used.

Figure 6. Methods to determine whether particles in an aerosol are biological in origin.

Parameter	Rationale	Test	Status
Virulence Determination (Many Virulence factors are contained in plasmid.)	A variety of pathogens and toxins bind to specific receptors such as gangliosides.	Binding to specific receptor	Available (Lab) – Approach has been used with cholera toxin, SEB, and others in a variety of formats
	Many virulence factors associated with bacterial plasmids	Plasmid determinations by Flow cytometry	Demonstrated in laboratory.
		Immunoassay for plasmid products.	Demonstrated in laboratory.
	Binding to a sentinel cell can be measured.	Live cell assays.	Demonstrated in the laboratory.

Figure 7. Methods to determine pathogenicity or virulence factors.

2.3. Pathogenicity/Virulence Determination

If a change in the bioflux is determined to be significant, the next step would be to determine whether this change is a biological material that is dangerous, i.e., may cause death or poses a threat to health. Thus we need to make a determination as to whether or not the material is a pathogen or a non-pathogen.

Various approaches can be used to make such a virulence determination. For example, many pathogenic materials bind to gangliosides, and this has been used as the basis for assays of several toxin materials for close to thirty years. More recently, DNA fragments, called aptamers, have been described, that can potentially be made to recognize specific pathogenic structures. Nucleic acid analysis may also be performed to detect specific nucleic acid sequences that could code for pathogenic markers. Siderophores have also been proposed to be used in this context. Lastly, assays can be performed for products of virulence plasmids, such as the determination of the PA component of *B. anthracis* toxin or some of the various YOP proteins of *Y. pestis*. Examples of various test methodologies are shown in Fig. 7.

Parameter	Rationale	Test	Status
Bacterial-Specific Fluorescence or Luminescence	Can selectively lyse sample with bacterial specific detergents, enzymes, phages, etc.	ATP Luminescence	Commercially Available
	There is a generic test for E.coli that could be extended to other enterics. (Specific substrates could also be found for other bugs.)	4-MUG Fluorescence	Available – Widely used for water monitoring.

Figure 8. Examples of generic bacterial detection using fluorescent techniques.

2.4. Classification of Agent

The final step in this detection process would be to classify the threat material as a bacteria, virus, proteinaceous toxin, or non-proteinaceous toxin. The most expedient way to make this determination could be through mass spectrometry.

Pyrolysis mass-spectroscopy does have this capability, but there are some alternative methods that can be used. Bacteria can be determined through the use of a Gram stain. Specific enzyme activities may also prove useful. For example, beta-galactosidase is an enzyme that is widely used to detect and presumptively identify the presence of E. coli in water samples.

Some other bacteria also possess similar specific enzymes. Figure 8 shows several approaches for generic bacterial detection/classification.

In the context that these bacteria are used as weapons, one may assume that antibiotic resistance has been introduced into them. There are various tests that are currently available clinically that use a colorigenic substrate that measures an antibiotic lytic enzyme, such as penicillinase.

More recent techniques utilize nucleic acid probes to measure the presence of DNA sequences in plasmids or plasmid constructs that code for these lytic enzymes.

A similar approach can be taken to determine the presence of virus. For example, some viruses possess specific enzyme activities that can be measured. The neuraminidase of Influenza virus is such an example. The properties of this enzyme were studied and were exploited in a rapid assay for detection.

Since viruses are intracellular parasites, one may be able to assume that there would be carrier cells or culture components present concurrently with the virus. One might be able to use an assay for ovalbumin in cases where eggs are used as the carrier. If conventional cell culture is used as the means to grow the virus, the mitochondria from them might be measurable by using a fluorescent dye, such as Rhodamine 123.

Histocompatibility antigens, which are species-specific antigens that are present on cell surfaces, may also prove useful as a means to detect the presence of viruses. These approaches are summarized in Fig. 9.

Proteinaceous toxins can be determined by several means. Conventional protein determination approaches, such as the Biuret, Coomassie Blue, and others, could serve as an initial screen. This could be followed by more stringent analysis, such as capillary electrophoresis and sequencing. This sequence could then be introduced into a bioinformatics tool, and a possible function could be determined. In the event that the toxin may have some type of enzyme activity associated with it, substrates for the enzyme can be determined and used in subsequent analyses. Figure 10 itemizes some approaches from proteinaceous toxin detection.

Detection of non-proteinaceous toxins would follow a similar approach. Detection could be based on virulence determination, mass-to-charge rations, or enzymatic activity, as shown in Fig. 11.

Parameter	Rationale	Test	Status
Viral-Specific Fluorescence or Luminescence	Neuraminidase is for Influenza detection.	Measure enzymes present on virus.	Demonstrated in the laboratory.
RNA Detection	Most BW viruses contain RNA. RNA content in the environment is low due to lack of active metabolism	Ribo Green Dye fluorescence by flow cytometry or simple fluorometry.	Dye available and demonstrated in the laboratory. Concept demonstrated with marine samples.
Carrier Cells	Viruses would probably have cells available as vectors.	Histocompatibility Antigens by Immunoassay.	Available (Lab) – demonstrated in laboratory.
		Mitochondrial detection using Rhodamine 123.	Available (Lab) – demonstrated in laboratory.

Figure 9. Approaches to viral detection/classification.

Parameter	Rationale	Test	Status
Protein Content	Protein content in air is low. Could be used as an additive or agent itself.	Standard Protein Determinations	Available (Field) – Various protection kits are routinely used
Virulence Determination	A variety of toxins bind to specific cell receptors and modulate the cell's chemistry.	Binding to receptor	Available (Lab) – Approach has been used with cholera toxin, SEB, and others in a variety of formats. Could be integrated into a bio-chip.
	Binding to a sentinel cell can be measured.	Sentinel Cell Assay	Demonstrated in the laboratory.
Marker Determination Sequence Determination	Scope of potential physiological active peptides is great.	Mass Spectrometry	Demonstrated in the laboratory.
Enzymatic Activity	A variety of toxins are either enzymes themselves or inhibit specific enzymatic processes.	Fluorogenic Substrate Analyses.	Demonstrated in the laboratory.

Figure 10. Methods to characterize protein toxins.

Parameter	Rationale	Test	Status
Virulence Determination	A variety of toxins bind to receptors.	Binding to receptor	Available (Lab) – Approach could be integrated into a bio-chip.
	Binding to a sentinel cell can be measured.	Sentinel Cell Assay	Concept demonstrated in the laboratory.
Molecular Weight Determination	Scope of potential physiological active materials is great.	Mass Spectrometry	Demonstrated in the laboratory.
Enzymatic Activity	Substrate Analyses.	A variety of toxins inhibit specific enzymatic processes.	Demonstrated in the laboratory.

Figure 11. Methods to characterize non-proteinaceous toxins.

Some of the approaches described in the virulence determination section can also be used to determine if the toxin is dangerous to life and health. For example, the Gm1 ganglioside is found on many cells and many of the pathogenic toxins bind to it. Specific examples include SEB and cholera toxin.

Parameter	Rationale	Test	Status
Cellular	Routine test for bacteriological identification	Metabolic fingerprint.	Available (Lab) – Possibly used in field.
	Routine test for strep throat	Immunoassays.	Available (Field)
	Virulence plasmids can be identified by nucleic acid hybridization techniques.	Nucleic Acid Analysis.	Available (Field) – Fieldable nucleic acid analysis system demonstrated on several platforms.
Specification Techniques: Indigenous vs non-indigenous strains	Accepted techniques for bacterial identification	Chromosomal DNA analysis	Available (Lab) – can be implemented into field situations.
		RNA fingerprinting	Available (Lab) – can be implemented into field situations.
Genetic Manipulation	Identify vector sequences. Identify markers associated with environmental stability.	Nucleic analysis	Available (Lab) – Needs additional development to expand available library.

Figure 12. Methods for Identification of biological agents.

From these approaches, detectors with improved capabilities to detect and warn may be developed that improve one's abilities to protect both troops and assets. Some of these technologies exist today and can be implemented into the field with some success. However, with the improvements in coating technologies and the integration of biological polymers with electronics, the next 50 years should see the development of detectors that utilize several of these approaches in concert. The use of chip technology, although now in its infancy, will play a crucial role in the future in these determinations. Since these technologies are widely known and have been summarized in various reviews, they will not be discussed here. A summary of identification techniques can be found in Fig. 12.

3. Identification and Medical Countermeasures

3.1. Identification

Once a determination has been made in the field that a biological event has occurred, the next step would be to retrieve the suspect samples and return them to a lab for further identification and classification. There are a plethora of techniques that are available for identification, such as metabolic tests, carbohydrate or fatty acid analysis, phage typing, immunological assays, and nucleic acid probe technologies. Current identification techniques use rigorous analysis to identify microorganisms according to complex taxonomic schemes, using both DNA and RNA analysis. The degree of relatedness among genus, species, and strains can thus be determined.

3.2. Medical Countermeasures

3.2.1. Rapid Susceptibility Testing

A current rationale, that one needs to know the identity of the particular agent so that the proper treatment modality can be employed, needs to be re-evaluated. In classical medical approaches, when one knows the identity of the organism, one can typically prescribe the appropriate course of antibiotics or other therapies. However, with the advent of genetic engineering and the relative ease that this allows an adversary to impart resistance to antibiotics, one can no longer assume that the mere identification of the organism would be sufficient for treatment. Even in conventional medical treatment, there has been an increase in the use of susceptibility tests to determine the appropriate course of treatment. In the case of an intentional release of a biological agent by an adversary, the use of antibiotic susceptibility testing is crucial.

In the event that a biological attack has occurred, particularly with a bacteria or a toxin, a viable sample is crucial for initiation of medical countermeasures. Although Koch's postulates will have to be demonstrated for legal purposes, the viable sample will be necessary so that antibiotic susceptibility testing can be performed. Currently these tests involve isolating the organism and eventually obtaining it in pure culture for further analysis. In the case of a bacterium, the Kirby-Bauer procedure is usually used and takes 8 or more hours to complete. In the event of a biological attack, there may not be sufficient time to do this. What is needed is a rapid method to do these determinations on environmental samples.

Several approaches have been proposed to approach this. One utilizes flow cytometry and rapid determinations in 20 minutes have been achieved. This approach uses a classical approach where the viable cells are mixed with an antibiotic mixture. The effect of the antibiotic on the cells is then measured by changes in scatter or DNA-specific dye uptake. Alternatively, analysis with nucleic acid probes that are specific for sequences that code for antibiotic resistancehave been proposed. This has the added advantage over conventional techniques in that viable samples are not required.

A similar paradigm can be assumed for viruses, but toxins are a different case. Specific identification needs to be obtained so that the proper antidotes, if available, can be administered. In the case of real unknown entities, bio-informatics will eventually play a role so that the possible physiological activity of the material can be obtained. However, we are several years away from this being a field consideration.

3.2.2. Immune Status Determinations

Up until this point, the agents themselves have been discussed; however, the other part of the equation concerns the troops that may have been attacked. It may be desirous to make a determination as to who has been attacked so that the field commander may not have to compromise his military posture. One possibility is to measure the status of immune function of the individuals who were involved in the attack. There are several approaches that may be used to assess immune status, from the classical blood cell counts to measuring individual lymphocyte status by flow cytometry analysis. As our ability to measure these functions improves, they may prove to be viable assets in the field.

4. Deployment Strategies

The previous discussion dealt with the types of technologies that one could conceivably use to detect the presence of biological materials; however, concepts of employment still need to be determined. This is the difficult part because it is here where the considerations are more cost/benefit, logistics, and personnel driven, rather than science driven. If we were to determine the types of scenarios that one would get in a potential terrorist scenario, there would be 2 cases: high value, fixed assets (buildings, stadiums, etc.) and hoax scenarios. In both of these cases, the scenarios are quite different. Another consideration is responding post attack. Here we need to determine of areas are contaminated and then perform quality control on personnel and materiel decontamination.

If there are some high value, fixed assets where a threat is credible, then some type of continuous monitoring system is probably worthwhile considering. These types of systems could utilise one or more of the triggering/detection technologies. For example, particle size and shape analysis coupled with fluorescence, could be an effective system. Systems such as the CDC 4WARN or the BAWS system under development in the US could be likely candidates. To minimise false alarms, this approach could be coupled with a detector system that utilises some of the detector schemes, such as bio/non-bio or pathogen/non-pathogen. An ideal candidate for this type of approach is a flow cytometer or similar device whereby several parameters can be coupled together in one platform. However, the down side of this approach is that it would require some degree of maintenance by building personnel.

Hoax scenarios can be dealt with by relatively simple equipment. One initially needs to make a bio/non-bio determination by a DNA or protein test. If positive, a viability test could then follow to determine if the sample contains live organisms. Samples can then be further analysed by identification technologies, either on or off site.

Responding to an attack, however, requires an echelon of response. One first needs to determine if there is indeed a biological agent present. This can be determined by the same means discussed in the hoax scenario. If it is determined that indeed live biological material is used, then the area can be sampled to determine the extent of contamination. Lastly, the area can then be decontaminated and the effectiveness of this process can be determined by several simple tests, such as luminescence.

5. Conclusions

From this brief discussion, it can be shown that the problem is not insurmountable; however, several things need to change. We first need to become aware that the possibility does indeed exist that "unconventional" biological agents be encountered in the field. There are a variety of strategies that could be implemented in either trigger or detection platforms that could be used to detect signatures from biological agents and possibly determine that they could present a danger to health and life.

However, we need to change our paradigms on how we think about the problem. The most important thing is to provide a detection and warning system that exploits credible signatures so that those in peril can take the appropriate protective measures.

References

Adam P. Flame photometry for biological detection. Proceedings of the Sixth International Symposium on Protection Against Chemical and Biological Warfare Agents, 1998, 61–67.

Bartoszcze M, Bielawska A. The Past, Present, and Future of Luminometric Methods in Biological Detection, 73–78.

Boulet CA, Hung G, et al. Capillary Electrophoresis/Nucleic Acid Probe identification of Biological Warfare Agent Simulants, 87–92.

Bruno JG, Mayo MW. A color image analysis method for assessment of germination based on differential fluorescence staining of bacterial spores and vegetative cells using acridine orange. Biotechnic and Histochemistry 1995; 70(4): 175–184.

Bruno JG, Yu H, et al. Development of an immunomagnetic assay system for rapid detection of bacteria and leukocytes in body fluids. J. Mol Recog 1996; 9: 474–479.

Bruno JG, Kiel JL. In vitro selection of DNA aptamers to anthrax spores with electroluminescence detection. Biosensors and Bioelectronics 1999.

Bryden WA, Benson RC, et al. Tiny-TOF Spectrometer for BioDetection, 101–110.

Button DK, Robertson BR. Determination of DNA content in aquatic bacteria by flow cytometry. Appl. Envrion. Microbiol. 2001; 67(4): 1636–1645.

Darzynkiewicz Z, Bedner E, et al. Laser-Scanning cytometry: a new instrumentation with many applications. Experimental Cell Research 1999; 249: 1–12.

Davey HM, Kell DB. Flow cytometry and cell sorting of heterogeneous microbial populations. Microbiol. Reviews 1996; 60: 641–696.

Del Vecchio VG, Redkar R, et al. Development of PCR-Based assays for the detection and molecular genotyping of microorganisms of importance in biological warfare, 219–229.

Ekins RP. Immunoassay, DNA analysis, and other ligand binding assay techniques: From electropherograms to multiplexed, ultrasensitive microarrays on a chip. J Chem Ed 1999: 76(6): 769–788.

Ezzell JW, Abshire TG. Immunological analysis of cell associated antigens of *Bacillus anthracis*. Infect. Immun. 1988; 56: 349–356.

Ezzell JH, Abshire TG, et al. Identification of *Bacillus anthracis* by using monoclonal antibody to cell wall galactose-N-acetylglucosamine polysaccharide. J Clin Microbiol 1990; 28: 223–31.

Forsberg A. A shared strategy for virulence of bacterial pathogens. Proceedings of the Sixth International Symposium on Protection Against Chemical and Biological Warfare Agents, 1998, 163–172.

Garrigue H, Patra G, and Ramisse V. Use of PCR for Identification and Detction of Biological Agents, 259–278.

Hairston PP, Ho J, Quant FR. Design of an instrument for real-time detection of bioaerosols using simultaneous measurement of particle aerodynamic size and intrinsic fluorescence. J. Aerosol Sci. 1997; 28(3): 471–482.

Ho J, Spence M, and Hairston P. Measurement of Biological Aerosol with a Fluorescent Aerodynamic Particle Sizer (FLAPS): Correlation of Optical Data with Biological Data, 177–201.

LeBaron P, Servais P, Agogue H, Courties C, Joux F. Does the nucleic acid content of individual bacterial cells allow us to discriminate between active cells and inactive cells in aquatic systems? Appl. Environ. Microbiol. 2001;67(4): 1775–1782.

Mason DJ, Shanmuganathan S, et al. A fluorescent gram stain for flow cytometry and epifluorescence microscopy. Appl environ Microbiol 1998: 64(7): 2681–2685.

Maltsev VP, Cheruysvev V. Method and device for determination of parameters of individual microparticles. US Patent Number 5,560, 847, issued 22 July 1997.

Olive DM, Bean P. Principles and applications of methods for DNA-based typing of microbial organisms. J Clin Microbiol 1999; 37(6): 1661–1669.

Robertson BR, Button DK, Koch AL. Determination of the biomasses of small bacteria at low concentrations in a mixture of species with forward light scatter measurements by flow cytometry. Appl Environ Microbiol 1998; 64(10): 3900–3909.

Rolland X. Chemscan™ RDI: A real time and ultra-sensitive laser scanning cytometer for microbiology. Applications to water, air, surface, and personnel monitoring. Proceedings of the Sixth International Symposium on Protection Against Chemical and Biological Warfare Agents, 1998, 103–110.

Rowe CA, Tender LM, et al. Array biosensor for simultaneous identification of bacterial, viral, and protein analytes. Analytical Chemistry 1999: 71(17); 3846–3852.

Sincock SA, Kulaga H, et al. Applications of flow cytometry for the detection and characterization of biological aerosols. Field Anal Chem Tech 1999:3:291–306.

Stopa PJ Tieman D, et al. Detection of biological aerosols by luminescence techniques. Field Anal Chem Tech 1999:3:283–290.

Stopa PJ and Bartoszcze MA. Rapid Methods for Analysis of Biological Materials in the Environment. NATO ASI Series, KluwerAcademic Publishers, Dordrecht, NL, 2000.

Stopa PJ. The flow cytometry of *Bacillus anthracis* spores revisited. Cytometry 2000; 41(4): 237–244.

Walberg M, Gaustad P. Steen HB. Rapid assessment of ceftazidine, ciprofloxacin, and gentamicin susceptibility in exponentially-growing *E. coli* cells by means of flow cytometry. Cytometry 1997; 27: 169–178.

Walt DR, Franz DR. Biological Warfare Detection. Analytical Chemistry 2000; December 738A–746A.

Wiener SL. Strategies for the prevention of a successful biological warfare aerosol attack. Military Medicine 1996; 161(5): 251–256.

Zubkov MV, Fuchs BM, et al. Determination of total protein content of bacterial cells by SYPRO staining and flow cytometry. Appl Environ Microbiol 1999; 65(7): 3251–3257.

Emerging Biological Threat
G. Berencsi et al. (Eds.)
IOS Press, 2005

Clinical Aspects of Bioterrorism (Anthrax, Plague and Smallpox)

László RÓKUSZ, MD
Central Military Hospital of Hungarian Defense Forces, Budapest

The consideration of biological warfare has moved from the conventional military theatre to a terrorist operated bioterrorism event that's puts millions of people in cities across the globe under the threat of artificially acquired, life threatening infectious disease scenario. Biological warfare diseases are likely to present as one of a limited number of clinical syndromes. Plague, Staphylococcal Enterotoxin B and tularemia may present as pneumonia. Unfortunately many biological warfare diseases (Venezuelan Equine encephalitis, Q-fever, Brucellosis) may present as fever of unknown origin (FUO). Moreover, other diseases (anthrax, plague, tularaemia, smallpox) have undifferentiated febrile prodromes. Physicians must be able to identify early victims and recognize patterns of disease. The speaker will cover the clinical aspects of three most important infectious diseases caused by biological warfare agents: anthrax, plague and smallpox.

Recognition of need for local, regional, and national preparedness for threat against bioterrorism provides an opportunity to enhance the public health system and its linkages with current and new partners.

The consideration of biological warfare has moved from the conventional military theatre to a terrorist managed bioterrorism threat that's puts millions of people in cities across the globe under the threat of artificially acquired, life threatening infectious diseases. Among weapons of mass destructions, biological weapons are more destructive than chemical weapons, including nerve gas, and nuclear weapon (a few kilograms of anthrax can kill as many people as a Hiroshima-size nuclear weapon). Biological and chemical warfare diseases can generally divided into two parts: 1. that present acutely with little or no incubation or latent period (usually the chemical agents) and 2. with a considerable delay in presentation (usually the biological threat agents). Biological warfare diseases are likely to present as one of a limited number of clinical syndromes. Plague, Staphylococcal Enterotoxin B and tularemia may present as pneumonia. Unfortunately many biological warfare diseases (Venezuelan Equine encephalitis, Q-fever, Brucellosis) may present as fever of unknown origin (FUO). Moreover, other diseases (anthrax, plague, tularaemia, smallpox) have undifferentiated febrile prodromes. Physicians must be able to identify early victims and recognize early patterns of diseases. The speaker will cover the clinical aspects of three most important infectious diseases caused by biological warfare agents: anthrax, plague and smallpox.

1. Anthrax

Anthrax is a zoonotic disease, caused by *Bacillus anthracis*.

1.1. Etiology

B. anthracis is a Gram-positive, aerobic, spore-forming species. Sporulation occurs under adverse environmental conditions and when vegetative bacteria are exposed to air. The spores are extremely hardy and can survive extremes of temperature, dryness, and flooding. When conditions improve, the spores germinate to produce vegetative bacteria.

1.2. Pathogenesis

Bacillus anthracis produces a toxin consisting of three components: edema factor, lethal factor, and protective antigen – that combine to form 2 toxins: lethal toxin and edema toxin. The protective antigen allows the binding of lethal and edema factors to the affected cell membrane and facilitates their subsequent transport across the cell membrane. Edema toxin impairs neutrophil function in vivo and affects water homeostasis leading to edema, and lethal toxin causes release of TNF-α and interleukin-1ß, factors that are believed to be linked to the sudden death in severe anthrax infection [1]. Anthrax infection occurs in humans by 3 major routes: inhalational, cutaneous, and gastrointestinal.

1.3. Clinical Manifestations

Anthrax has 4 clinical forms: a) cutaneous b) gastrointestinal c) inhalational d) meningeal. A biological warfare attack with anthrax spores delivered by aerosol would cause inhalational anthrax.

Cutaneous anthrax (malignant pustule) usually occurs on the hands and forearms or face [2]. After an incubation period of three to ten days a small pruritic, painless papule develops at the site of inoculation. Several days later a small vesicle is noted. The vesicle typically dries and black eschar develops from it. Significant edema may develop. There may be mild fever, malaise, and headache. The local infection can occasionally disseminate into a severe systemic infection. The fatality rate in untreated cutaneous anthrax is up to 20%, but with early and effective therapy it can be reduced to less than 5%.

Gastrointestinal anthrax occurs following the ingestion of anthrax spores. Significant abdominal pain, hematemesis, and bloody diarrhoea may develop.

Inhalational anthrax occurs when individuals working with animal wool inhale the spores. Also, inhalational anthrax may occur from inhalation of aerosolised spores released during a biological warfare attack. The incubation period for anthrax is hours to 7 days. In Sverdlovsk, cases occurred from 2 to 43 days after exposure. This form of the disease is biphasic. Early signs and symptoms suggest an influenza-like illness, with mild fever, malaise, fatigue, myalgia, non-productive cough, nausea or vomiting, chest discomfort and pleural chest pain. Within several days after infection, there may be clinical improvement. The next phase is heralded by progressive respiratory signs and symptoms. Patients have dyspnea, cough, respiratory stridor, profuse diaphoresis. Minimal pleural effusion may be evident and shock may develop. On chest x-ray, the typical finding is that of an enlarged mediastinum, and possible pleural effusion. Sepsis syndrome and hemorrhagic meningitis may be present.

1.4. Laboratory Findings

Laboratory studies may reveal leukocytosis, with left shift, hypoxemia, and elevated transaminases. The patient usually dies within 24 hours.

During the early phase, blood and respiratory secretions may be sent for rapid identification by polymerase chain reaction (PCR). A rapid diagnostic test is available for detec-

Table 1. Recommended Therapy for Inhalational Anthrax infection.

Category	Recommended therapy	Duration
Adults	**Ciprofloxacin 2x400 mg/d iv. or** Doxycyclin 2x100 mg/d iv. and 1 or 2 additional antibiotics[1] iv. Ciprofloxacin 2x500 mg/d po. Or Doxycyclin 2x100 mg/d po.	60 days
Children	**Ciprofloxacin 2x10 mg/kg iv. or** Doxycyclin 2x100 mg iv. (if weight >45 kg) or 2x2,2 mg/kg iv. (if weight < 45 kg) Ciprofloxacin 2x15 mg/kg po. or Doxycyclin 2x100 mg po. (if weight >45 kg) or 2x2,2 mg/kg po. (if weight < 45 kg)	60 days
Pregnant women	Same for nonpregnant adults	60 days
Immuno-compomised Persons	Same for immunocompromised persons	60 days

[1] Other agents in vitro activity include penicillin, ampicillin, imipenem, clindamycin, chloramphenicol, clarithromycin, rifampin and vancomycin.
[2] Iv. treatment initially before switching to oral antibiotic therapy when clinically appropriate.

tion of toxin antigens in the blood during acute phase. Serologic identification of *B. anthracis* isolates by a protective antigen-specific monoclonal antibody dot-ELISA has recently been described [3].

Chest x-ray may appear normal or show hilar adenopathy, and widened mediastinum and pleural effusions during the acute phase. Chest CT scan will show hyperdense hilar and mediastinal nodes, mediastinal edema, infiltrates, and pleural effusions.

1.5. Treatment

The anthrax attacks of 2001 in the USA resulted in 11 cases of inhalational anthrax, 5 of which turned fatal [1]. Given the rarity of anthrax infection, the first clinical or laboratory suspicion of an anthrax disease must lead to early initiation of antibiotic treatment awaiting confirmed diagnosis and should stimulate immediate notification of the local or state (county) public health department, local hospital epidemiologist, and local or state (county) public health microbiological laboratory. Two or three antibiotic combination treatment may have therapeutic advantage (early treatment with fluoroquinolone with doxycycline or clindamycin or macrolide or penicillin). Some ID experts have also advocated the use of clindamycin, because of theoretical benefit of diminishing bacterial toxin production. Some ID specialists recommended preferential use of ciprofloxacin over doxycycline with chloramphenicol; rifampicin or penicillin, when meningitis is suspected or established. For asymptomatic patients with suspected exposure to anthrax spores can be achieved with a 6-week course or doxycyclin or ciprofloxacin. Standard anthrax vaccine in the US is approved by the FDA and is routinely administered to persons at risk for exposure to anthrax spores.

In the Table 1 and 2 there are shown the CDC summary for the treatment and postexposure prophylaxis of inhalational anthrax [1].

Supportive care includes maintaining the airway, providing fluids, and vasopressors as indicated for shock. A tracheostomy is indicated for upper airway obstruction. Surgical de-

Table 2. Recommended Therapy for Inhalational Anthrax infection in the Mass Casualty setting or for Post-exposure Prophylaxis.

Category	Recommended therapy	Duration
Adults	Ciprofloxacin 2x500 mg/d po. or Doxycyclin 2x100 mg/d po. or Amoxicillin 3x500 mg/d po.	60 days
Children	Ciprofloxacin 2x15 mg/kg po. or Amoxicillin 3x500 mg/d po. (if weight > 20 kg) 3x80 mg/kg po. (if weight < 20 kg)	60 days
Pregnant women	Ciprofloxacin 2x500 mg/d po. or Amoxicillin 3x500 mg/d po.	60 days
Immunocompomised Persons	Same for immunocompromised persons	60 days

bridement of cutaneous lesions is contraindicated, because of possibility of dissemination. Surgical drainage of mediastinum for inhalational anthrax is nor recommended.

1.6. Prophylaxis

Anthrax vaccine (AVA=anthrax vaccine adsorbed) is an inactivated cell-free product, is given in six doses at 0, 2, and 4 weeks and 6, 12, and 18 months, with annual boosting.
Contraindications to vaccination include:

1. History of anaphylaxis after receiving a dose of AVA or any of the vaccine components
2. History of anthrax infection [1,4].

2. Plague

Plague is a systemic zoonotic infection caused by *Yersinia pestis*. The pathogen has been the cause of 3 great pandemics in the common era: in the 6th, 14th (the Black Death pandemic), and 20th centuries. During the Vietnam War, plague was endemic among the native population, but U.S. soldiers escaped relatively unaffected (only 8 American troops were affected) [5]. American success was attributed to the use of flea insecticide, immunization of virtually all American troops with plague vaccine, and use of insect repellents, protective clothing, and rat-proofed dwellings. *Y. pestis* is one of the most serious biological weapon, that that could cause disease and death in sufficient numbers to cripple a city or region. In 1970, the World Health Organization (WHO) reported that, in a worst-case scenario, if 50 kg of *Y. pestis* were releasedas an aerosol over a city of 5 million, pneumonic plague could occur in as many as 150.000 persons, 36.000 of whom would to die [6].

2.1. Etiology

Y. pestis is a Gram-negative, nonmmotile, non-sporulating, aerobic coccobacillus, that shows bipolar staining with Wright, Giemsa, or Wayson stain. The organism grows readily on most culture media, ordinary nutrient agar, blood agar, or deoxycholate agar. Y. pestis produce a number of antigenic components, several of which are of major importance in the pathogenesis of disease. These include a capsular antigen, which is heat-labile protein (Fraction 1; F1), and the VW antigens, which include a protein V and lipoprotein W. The F1 and VW antigens which render the organism resistant to phagocytosis.

Y. pestis contains an endotoxin (may be less lethal), and exotoxins, which are lethal (cardiotoxic) for the mouse and rat.

2.2. Epidemiology

Human plague most commonly occurs when plague-infected fleas (*Xenopsylla cheopis*) bite humans who then develop bubonic plague. The bubonic form may progress to the septicemic and/or pneumonic form. Pneumonic plague would be the predominant form, after aerosol dissemination. The black rat, *Rattus rattus*, has been most responsible worldwide for the persistence and spread of plague in urban epidemics. The greatest risk to humans occurs when large concentrations of people live under unsanitary conditions in close contact. Respiratory droplet transmission can occur person-to-person. The incubation period may vary from a few hours to 12 days, but is usually 2–5 days.

2.3. Pathogenesis

Bite by plague infected flea causes inoculation of organisms, after that *Y. pestis* penetrate into the regional lymph nodes, after that necrosis develops, and causes bacteremia and dissemination (pneumonia, meningitis sepsis) and shock, DIC and coma.

2.4. Clinical Manifestations

Plague have 3 clinical forms: a) bubonic b) pneumonic and c) septic.

The onset is abrupt and often associated with chills, and the temperature rises to 39–41°C. Symptoms include malaise, headache, vomiting, chills, altered mentation, cough, abdominal pain, chest pain. Few hours after onset of symptoms, buboes with pain can occur. A skin lesion at the portal of entry is seen in less than 25% of cases. Tachycardia, hypotension, leukocytosis, and fever are frequent symptoms to be developed. Approximately 5–15% of bubonic plague patients will develop secondary pneumonic plague, due to hematogenous dissemination. The case fatality rate for untreated bubonic plague is about 60%, but less than 5% with prompt, effective therapy.

Primary pneumonic plague has an incubation period usually lasting only 2–3 days. Initially cough is not prominent symptom. Within 24 hours a productive cough develops. Sputum is mucoid or blood-tinged. There is associated dyspnoe and tachypnoe, rales may be absent. Chest X-ray shows a rapidly progressing pneumonia.

Septicemic plague may evolve from any form of plague, presents as an acute fulminant illness which may be fatal before either lymph node or pulmonary manifestations predominate.

2.5. Laboratory Findings

Laboratory findings include leukocytosis, with a total WBC count up to 20.000 cells/mm^3 with increased bands, and greater than 80% neutrophils. When DIC is present, fibrin degradation products will be elevated, and thrombocytopenia develops. Because of liver involvement, alanine aminotransferase, aspartate aminotransferase, and bilirubin levels are often increased.

Table 3. Recommended Therapy for patients with Pneumonic Plague.

Category	Recommended therapy	Duration
Adults	**Streptomycin inj. 2x1 g/d im. or** Gentamicin inj. 1x5 mg/kg/d im. or iv. or 2 mg/kg loading dose followed by 3x1,7 mg/kg im. or iv.[1] or Ciprofloxacin inj. 2x400 mg/d iv. or Doxycyclin inj. 2x100 mg/d iv. or Chloramphenicol inj. 4x25 mg/kg iv.[2]	10–14 days
Children	**Streptomycin inj. 2x15 mg/kg im. (maximum daily dose 2 g)** Gentamicin inj. 3x2,5 mg/kg im. or iv.[1] Doxycyclin inj. 2x100 mg iv. (if weight >45 kg) or 2x2,2 mg/kg iv. (maximum 200 mg/d) (if weight < 45 kg) Ciprofloxacin inj. 2x15 mg/kg po. Or Chloramphenicol inj. 4x25 mg/kg	10–14 days
Pregnant women	Gentamicin inj. 1x5 mg/kg/d im. or iv. or 2 mg/kg loading dose followed by 3x1,7 mg/kg im. or iv.[1] or Ciprofloxacin inj. 2x400 mg/d iv. Or Doxycyclin inj. 2x100 mg/d iv.	10—14 days

[1] Aminoglycosides must be adjusted to renal function.
[2] For the treatment of plague meningitis.

A presumptive diagnosis can be made microscopically by identification of the coccobacillus in Gram, Wright, Giemsa or Wayson stained smears from lymph node needled aspirate, sputum, blood, or cerebrospinal fluid samples or periferic blood may disclose gram-negative bacilli. Immunofluorescent staining is very useful. Cultures of sputum, blood, lymph node aspirate should demonstrate growth 24–48 hours after inoculation. Serologic tests include passive hemagglutination and complement fixation.

2.6. Treatment

Suspected pneumonic plague cases require strict isolation with Droplet Precautions for at least 48 hours of antibiotic therapy.

Streptomycin, gentamicin, doxycyclin, ciprofloxacin, chloramphenicol are highly effective. Patients presenting after 24 hours following the onset respiratory symptoms less likely to survive. No vaccine currently available for prophylaxis of plague. Face-to-face (within 2 meters) contacts of patients pneumonic plague or persons possibly exposed to a plague aerosol should be given antibiotic prophylaxis for 7 days or the duration of risk of exposure plus 7 days (doxycyline or ciprofloxacin). In the Table 3 the CDC summary for the treatment of patients with pneumonic plague are shown [7].

2.7. Prophylaxis

2.7.1. Vaccine

No vaccine is currently available for prophylaxis of plague.

2.7.2. Antibiotics

Face-to-face contact of patients with pneumonic plague or persons possibly exposed to a plague aerosol in a plague biological weapon attack should be given antibiotic prophylaxis for 7 days. The choice of antibiotic for chemoprophylaxis is doxycycline (2x100 mg orally). Ciprofloxacin (2x500 mg daily orally) has also shown to be effective.

Contacts of bubonic plague patients need only be observed for symptoms for 7 days. If symptoms develop, start antimicrobial therapy.

3. Smallpox (Variola)

The disease was declared to be eradicated on May 8, 1980. by WHO. Despite the global eradication of smallpox the potential weaponization of variola continues to pose a military threat.

3.1. Etiology

Smallpox is caused by the *Orthopox* virus. The poxviruses are the largest viruses so recognized. Smallpox, a DNA virus, is a member of the genus orthopoxvirus. Three other members of this genus (vaccinia, cowpox, and monkeypox) can also infect humans.

Variola viruses classified as variola major and minor (or alastrim) on the basis of the severity of clinical illness caused by infection.

3.2. Epidemiology

Through the end of the 19th century, variola major predominated throughout the world. Typical variola major epidemics resulted in fatality rates of 30% or higher among the unvaccinated. In the early 20th century, a milder clinical form of smallpox became prevalent in the Americas, Europe, and South Africa. Fatality rates were 1% or less [9].

Smallpox spreads from person to person, primarily by droplet nuclei or aerosols expelled from the oropharynx of infected persons and by direct contact. Contaminated clothing or bed linens can also spread the virus.

Before global eradication, the only reservoir for variola virus were humans. No natural reservoir for the virus currently exists. Stocks of variola virus have been retained in two WHO-approved collaborating centres: the Centers for Disease Control and Prevention (CDC) in Atlanta and the Russian State Centre for Research on Virology and Biotechnology (Koltsovo, Novosibirsk Region, Russian Federation).

The infectious dose is presumed to be low (10 to 100 organisms). The observed secondary attack rates among susceptible close contacts have varied from 37% to more than 70%. The average number of cases infected by a primary case is estimated at 3.5 to 6.

Smallpox is of concern as a biological weapon for several reasons: much of the population is susceptible to infection, the virus carries a high rate of morbidity and mortality, vaccine is not yet available for general use, and past experience has demonstrated that introduction of the virus creates a great deal of havoc and panic.

3.3. Pathogenesis

The site of entry of the smallpox virus is usually through the oropharyngeal or respiratory mucosa, but the variola virus also can enter through the skin, and rarely through the conjunctiva or placenta. The virus migrates rapidly to the regional lymph nodes. Asymptomatic viremia occurs on the 3rd or 4th day after inoculation, with further dissemination of the virus to spleen, other lymph nodes and bone marrow. Secondary viremia develops by the 8th to 12th day after infection, followed by onset of fever and toxemia. Clinical manifestations begin acutely with malaise, fever, vomiting, headache, backache. A total of 15% of patients develop delirium. Two to three days later, an enanthem appears concomitantly with a discrete rash on the face, hands, and forearms. Following subsequent eruptions on the lower extremities, the rash spreads centrally during the next week to the trunk. The virus multiplied in the epithelial cells of the skin and mucous membranes, causing pustulation. Lesions are more abundant on the extremities and face, and this centrifugal distribution is an important diagnostic feature. From 8 to 14 days after onset, the pustules form scabs, which leave depressed depigmented scars on healing. Scabs separated during the third to fourth week of illness and, although virus could be detected in these scabs [10].

3.4. Clinical Manifestations

The incubation period of smallpox is about 12 (range 7–17) days. The illness begin acutely with malaise, fever, rigors, vomiting, headache, and backache; 15% of patients developed delirium. Following the initial febrile period, a macular rash developed, occurring first on the mucosa of the mouth and pharynx, the face, and forearms, and then spreading to the trunk and legs. Distribution of the rash is centrifugal. The rash quickly became papular, and within two days the papules developed into vesicles and then pustules. Crusts begin to form on about the 8th or 9th day of rash. As the patient recovers, the scabs separate and characteristic pitted scarring gradually develops. Secondary bacterial infection is not common, and deaths which usually occurs during the second week of the smallpox, most likely results from the toxemia. The hemorrhagic form of smallpox characterized by a severely prostrating prodromal illness with high fever and head, back, and abdominal pain. The hemorrhagic variant was nearly always fatal. Death usually occurs by the 5th or 6th day after onset of rash.

The most frequent complications are: encephalitis (1:500), keratitis with corneal ulceration leading to blindness occurred in 1% of cases, pulmonary edema.

3.5. Differential Diagnosis

Smallpox must be distinguished from other vesicular exanthems, such as chickenpox, erythema multiforme, allergic contact dermatitis, impetigo, hand, foot and mouth disease, rickettsiosis, septicemic mieloidosis, autoimmune diseases (dermatitis herpetiformis).

3.6. Laboratory Findings

The usual method of diagnosis is demonstration of characteristic virions on electron microscopy of vesicular scrapings. The development of PCR techniques promises a more accurate method of microbiological diagnosis. A real-time PCR method based on the E9L gene has been developed, validated, and developed for vaccinia virus [9]. Under light microscopy, aggregation of variola virus particles, called Guarnieri bodies are found.

3.7. Treatment

There is no specific antiviral therapy available for smallpox. Medical personal provides supportive care: adequate fluid intake; alleviation of pain and fever; keeping the skin lesions clean to prevent bacterial superinfection.

Antiviral agents for use against smallpox are under investigation Cidofovir has been shown to have significant in vitro and in vitro activity in experimental animals. Cidofovir is a nucleosid analog DNA polymerase inhibitor, might prove useful in preventing smallpox infection if administered within 1 or 2 days after exposure [10]. But cidofovir must be administered iv. and its use is often accompanied by serious renal toxicity.

3.8. Infection Control

Any confirmed case of variola should be considered an international emergency with immediate report made to public health authorities. Droplet and Airborne Precautions for minimum 17 days following exposure for all persons in direct contact with the index case, especially the unvaccinated. Patients should be considered infectious until all scabs separate. Immediate vaccination or revaccination should also be undertaken for all personnel exposed to either weaponized smallpox virus or a clinical case of variola.

3.9. Prophylaxis

3.9.1. Vaccine

After declaring global eradication of smallpox, the WHO in 1980 recommended: "Smallpox vaccination should be discontinued in every country except for investigators at special risk."

Smallpox vaccine (vaccinia virus) is most of ten administered by intradermal inoculation with a bifurcated needle. Vaccination after exposure to weaponized smallpox or a case os variola may prevent or ameliorate disease if given as soon as possible and preferably within 7 days after exposure. A vesicle typically appears at the vaccination site 5–7 days after vaccination, with surrounding erythema and induration. The lesion forms a scab and heals over the next 1–2 weeks.

Side effects include: low-grade fever, axillary lymphadenopathy, postvaccinial encephalitis (occurred at a rate of 1 case per 300.000 vaccinations), progressive vaccinia, eczema vaccinatum, generalized vaccinia, inadvertent inoculation, idiosyncratic reactions (generalized urticarial exanthems, erythema multiforme, Stevens-Johnson syndrome.

Contraindications: four groups of persons are at special risk of complication:

1. persons with eczema or other forms of chronic dermatitis
2. patient with leukaemia, lymphoma, or other malignancies
3. pregnant women
4. those receiving immunosuppressive drugs.

3.9.2. Passive Immunoprophylaxis

Vaccinia immunglobulin (VIG) is generally indicated for treatment of complications to the vaccinia vaccine, and should be available when administering vaccine [11].

Smallpox vaccination of health care workers, military personnel, and some first responders has begun in the United States in 2002–2003 as one aspect of biopreparedness [12–14].

Recognition of need for local, regional, and national preparedness for bioterrorism provides an opportunity to enhance the public health system and its linkages with current and new partners.

References

[1] Inglesby TV, O'Toole T, Henderson DA, et al. Anthrax as a Biological Weapon, 2002. Updated Recommendations for Management. JAMA 2002; 287:2236–2252.

[2] USAMRIID's. Medical management of biological Casualties handbook. Fourth Edition. February 2001. Fort Detrick, Frederick, Maryland.

[3] Sastry KS, Tuteja U, Santhosh PK. Et al. Identification of Bacillus anthracis by a simple protective antigen-specific mAb dot-ELISA. J. Med. Microbiol. 2003; 52:47–9.

[4] Bioterrorism. Information & Resources. Anthrax Medical Summary. www.idsociety.org/bt.
[5] Reiley CG, Kates ED. The clinical spectrum of plague in Vietnam. Arch Intern Med 1970; 126 (12):990–4.
[6] *Health Aspects of Chemical and Biological Weapons*. Geneva, Switzerland: World Health Organization; 1970:98–109.
[7] Inglesby TV, Dennis DT, Henderson DA, et al. Plague as a Biological Weapon, JAMA 2000; 283:2281–2290.
[8] Bioterrorism. Information & Resources. Plague Medical Summary. www.idsociety.org/bt.
[9] Fenner F, Henderson DA, Arita I, et al. Smallpox and its Eradication. Geneva, Switzerland: World Health Organization. 1988:1460.
[10] Bioterrorism. Information & Resources. Smallpox Medical Summary. www.idsociety.org/bt.
[11] USAMRIID's: Medical Management of biological casualties handbook. Fourth Edition, 2001. Fort Detrick Frederick, Maryland.
[12] Fulginiti VA, Papier A, Lane JM, et al. Smallpox vaccination: A Review, Part I. Clin Infect Disease 2003; 37:241–50.
[13] Fulginiti VA, Papier A, Lane JM, et al. Smallpox vaccination: A Review, Part II. Clin Infect Disease 2003; 37:251–71.
[14] Sepkowitz KA. How Contagious Is Vaccinia? N Eng J Med 2003; 348,439–46.

Emerging Biological Threat
G. Berencsi et al. (Eds.)
IOS Press, 2005

Clinical Virologists and the Smallpox-Threat: A Comparison Between Germany and the UK

Joachim J. BUGERT
Department of Medical Microbiology, University of Wales College of Medicine, Heath Park, Cardiff, CF14 4XN, UK

In February 2003 the German Government bought 100 million doses of smallpox vaccine in response to an abstract threat of reemerging smallpox through a bioterrorist attack. The interaction between clinical virologists, US Armed Forces and Public Health authorities in Germany in this situation will be discussed using the University of Heidelberg Clinical Center as an example. The author has since moved to the UK and will compare approaches to the smallpox threat between Germany and the UK.

Emerging Biological Threat
G. Berencsi et al. (Eds.)
IOS Press, 2005

Genetic Attributes for Tracking Unconventional Microorganisms

Abdu F. AZAD
University of Maryland School of Medicine, Baltimore, MD 21201

The current list of known CDC defined select agents includes over 30 highly pathogenic organisms representing mostly viruses and bacteria. The listed organisms are among the most dangerous pathogens known to humans. The criteria used for their classification are based upon their virulence, short incubation period, high morbidity and mortality, low infectious dose, ease for spread, and survival under environmental conditions. A number of these organisms have been developed as biological weapons and field-tested with variable effectiveness in the past. However, recent advances in the field of microbial genetics and molecular manipulations now allow *de novo* synthesis and composition of new organisms with the multifactorial enhancement of their virulence, and survival.

The new features of genetically engineered and modified bacteria could include additional virulence genes, host target selectivity sequences, selection for increased stability and resistance to variable environmental conditions and resistance to existing and commonly used antibiotics. Drastic genetic and behavioral modifications of select agents not only increase the potential of these microorganisms as an agents of bioterrorism, it also masks their timely diagnosis. Development of such monstrosity using bacterial shell carrying a complex genome composed of intra-species virulence genes and powerful secretory systems are no longer a fiction.

How we identify packaged bacteria and diagnose those exposed will be the core of this presentation. To achieve this goal, molecular features of the CDC category A and B bacteria will be analysed to identify "signature" sequences, draw a means for their rapid diagnosis and biotracking, devise new avenues to curtail their mass production and dissemination and ultimately their global control. In addition, means by which pathogenic microorganisms could be tagged for future source identification and tracking will be discussed.

Emerging Biological Threat
G. Berencsi et al. (Eds.)
IOS Press, 2005

An Attempt to Characterize Some Soil and Health Relevant Bacteria by FT-IR Spectroscopy

Zdenek FILIP[1,*], Susanne HERRMANN[a] and Jaromir KUBAT[b]
[a]*Federal Environmental Agency, Langen Building, Paul Ehrlich Strasse 29, 63225 Langen, Germany*
[b]*Research Institute for Crop Production, Dept. of Soil Biology, Drnovska 507, 162 00 Prague 6, Czech Republic*

An increasing concern exists about health relevant bacteria that could be spread in soil and other environments by application of organic wastes or due to deliberate releases. Thus, there is a need for rapid and reliable detection of such agents and their differentiation from common soil bacteria. We made an attempt in using FT-IR spectroscopy for this purpose. Soil bacteria *Arthrobacter oxydans* and *Azotobacter chroococcum*, and facultatively pathogenic bacteria *Enterobacter agglomerans* and *Escherichia coli* delivered well distinguished IR spectra at a following waverange: 3300 cm^{-1} mainly for structural units of nucleic acids; 3000–2800 cm^{-1} for cell membrane fatty acids; 1800–1500 cm^{-1} for saturated esters and cell proteins; 1500 cm^{-1} for proteins, lipids and phosphoesters; 1200–800 cm^{-1} for glycopeptides and phosphate groups of nucleic acids constituents. Cell mass harvested at different phase of culture growth, and bacteria cultivated in differently composed nutrient solutions, delivered somewhat different spectra. If the bacteria were grown in a low nutrient broth and harvested uniformly, however, distinct differences in the FT-IR spectra appeared (i) between *A. oxydans* and *A. chroococcum*, and (ii) between these common soil bacteria and both *E. agglomerans* and *E. coli*. No spectral differences were obtained between the latter members of the Enterobacteriaceae family.

1. Introduction

Soil microorganisms account for the richest and highly complex natural microbial community. They play the main role in the biogeochemical cycling of carbon, nitrogen, phosphorus, sulfur and other bioelements, and in this way, contribute to the sustainability of life. Yet, there is also an increasing concern about the presence of pathogenic microorganisms in soil which could involve risk for human, animal and plant health. As reviewed recently, different organic wastes, and especially those generated by livestock and regularly applied onto land, usually contain pathogenic bacteria and viruses [1]. In addition, some concern might also appear about a deliberate spread of pathogens onto land, e.g. by terroristic activities. There is no doubt that several pathogenic microorganisms survive and remain infective in soil for a long time [2–4]. If pursuing a reliable risk assessment, rapid detection

[1] The research accomplishment site.
[*] Present address: Hegstrauch 7, 35463 Fernwald, Germany; zdenek.fillip@t-online.de.

of pathogenic microorganisms and their differentiation from common soil bacteria is necessary.

Usually, the soil microbial community analysis is based on traditional cultivation methods but still more also modern approaches such as extraction and estimation of indicators molecules are in use [5,6]. As soon as in the seventies we attempted to use infrared spectroscopy as a tool in characterizing complex biomass of soil microorganisms by fingerprint-like patterns of different cell structural units [7,8]. Later on, Fourier-transform infrared spectroscopy (FT-IR) has been introduced as a usefull and versatile technique in microbial ecology, clinical microbiology, and general biology [9–12]. Recently, this technique has been repeatedly used for identification of some clinically or food-spoilage relevant microorganisms [13–17]. The aim of our present investigations was to assess whether FT-IR spectra could be used as a tool in differentiating between common soil bacteria and some potentially pathogenic ones. For this reason we compared the spectra of the widely spread soil bacteria *Arthrobacter oxydans* and *Azotobacter chroococcum* with those of a common indicator bacterium *Escherichia coli* and *Enterobacter agglomerans* both of them members of the *Enterobacteriaceae* family, that includes several pathogens.

2. Material and Methods

2.1. Microorganisms and Culture Conditions

Strains of *A. chroococcum* (DSM 281), *E. agglomerans* (K 339), and *E. coli* (DSM) were obtained from the German Collection of Microorganisms and Cell Cultures in Braunschweig. *A. oxydans* was a soil isolate from Langen (Germany). Bacteria were cultivated in a Minimum Nurient Broth (MNB) containing 0.10 g NH_4Cl, 0.10 g $MgSO_4 \times 7\ H_2O$, 0.01 g Ca $Cl_2 \times 2\ H_2O$, 0.90 g K_2HPO_4, 0.10 KH_2PO_4, and 5 mL of a trace elements solution in 1 l deionized water (pH 7.2). After autoclaving at 121°C for 20 min, 40 ml glucose solution (40 mM), and 3 ml vitamine solution (sterile filtered) were added. The latter solution contained 2.5 mg biotin, 6.5 mg calcium-panthotenate, 25 mg nicotine-acidamide, 12.5 mg thiamine-hydrochloride, 12.5 mg para-aminobenzoate, and 62.5 mg pyridoxi-hydrochloride in 100 ml of deionized water. After 24 h at 30°C on a shaker, one ml of the cultures was transferred into 50 ml fresh MNB and cultivated again for 24 h. The final cultures containing about 1×10^8 colony forming units (CFU) $\times$ ml^{-1} were centrifuged at 10 000 × *g* at 4°C for 15 min to harvest cell biomass. After a three times repeated washing with deionised water and a subsequent centrifuging, the cell biomass was freeze dried and stored before use. To obtain a growth curve, the cultivation of individual bacteria in a fermenter was extended up to 238 h. In some MNB cultures, glucose was replaced by fructose, or ammonium chloride was omitted. For some examinations starving bacterial cells were prepared. These were obtained from a freshly harvested and washed biomass that was re-suspended in 50 ml deionised water and incubated 24 h at 30°C on a shaker.

2.2. FT-IR Spectroscopy

Between 1.5 and 2 mg of cell biomass was mixed with 300 mg KBr (Uvasol® Merck), and discs were pressed at 250 atm. They were examined in a Bruker IFS 48 FT-IR spectrophotometer over the waverange 4000–400 cm^{-1} (2.5–25 µm) with the resolution of 2 cm^{-1}. Before and during collection of spectra the system was purged by a pre-dried and CO_2-free air in order to avoid possible spectral obstacles from H_2O and CO_2 molecules. The digitised spectra were stored in an auxiliary Aspect 1000 computer (Bruker).

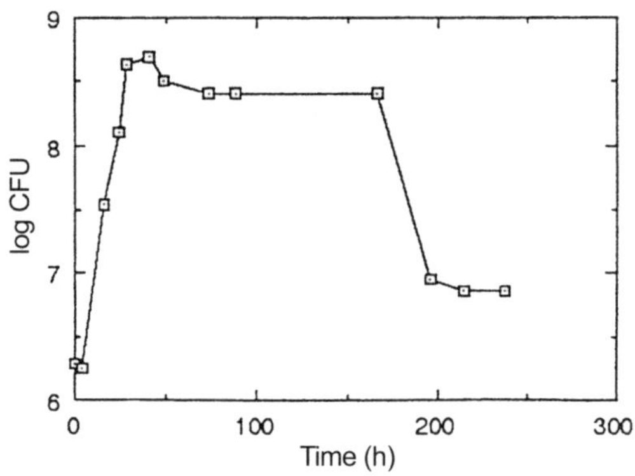

Figure 1. A growth curve of *A. oxydans* cultivated in a Minimum Nutrient Broth (MNB).

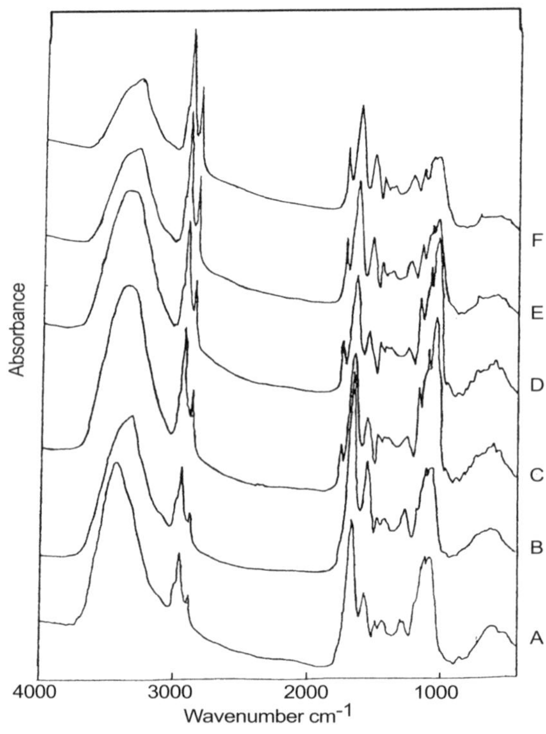

Figure 2. FT-IR spectra of *A. oxydans* cell mass harvested from MNB after 4 h (A); 16 h (B); 28 h (C); 48 h (D); 168 h (E); 238 h (F).

3. Results and Discussion

A. oxydans, a gram-positive autochthonous soil bacterium with a polymorphic rod-coccus life cycle, grew rather slowly in the MNB (Fig. 1). In the stationary phase of growth and thereafter, small spherically shaped cells dominated. FT-IR spectra in Fig. 2 (A-F) indicate distinct changes in the structural composition of the cell material in the course of bacterial

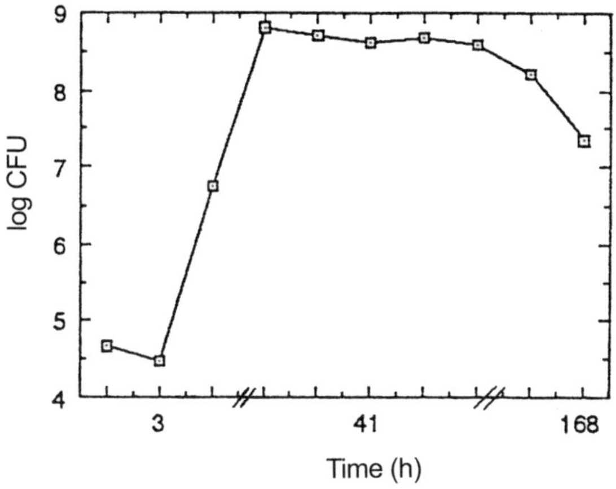

Figure 3. A growth curve of *E. coli* cultivated in MNB.

growth. Usually, a very strong IR absorption at 3400–3300 cm^{-1} should be attributed to the presence of moisture in KBr pellets. However, because precaution was made to suppress moisture (see in Section 2.2), and no free water band (1613 cm^{-1}) could be detected, rather some NH_2 stretching vibrations in adenine, guanine and/or cytosine structures of nucleic acids, and perhaps, OH groups adsorbed on DNA/RNA molecules should be made responsible for the strong absorption at the indicated waverange [18]. The intensity of this absorption an also of that at 1656 cm^{-1} (NH_2 bending, and C-N stretching vibrations in nucleic acids, amide I in proteins) increased during the exponential phase of growth and than decreased again. Simultaneously, the IR absorption in a region typical for aliphatic (fatty) acids at 3000–2800 cm^{-1} (-CH_2, -CH_3 groups) increased, and in addition, an absorption band at 1745 cm^{-1} (C-O in fatty acid esters) continuously developed. The raise of the absorption at 1745 cm^{-1} was positively correlated with values of ratios calculated from a relative intensity (a base to peak hight) of absorption bands at 2926 cm^{-1} and 1656 cm^{-1}. This value increased from 0.50 at early exponential phase of growth (after 4 h) to 1.45 at late stationary phase (after 168 h), and 1.62 at a phase of cell dying (after 238 h). These features indicate that the concentration of lipids in bacterial cell continuously increased in the course of culture ageing. An IR absorption triplet at 1150, 1080, and 1030 cm^{-1} that can be attributed to P-O-C and C-O-C stretching vibrations in glycopeptides, phosphoesters and saccharides, showed also quite dynamic developments in the course of bacterial growth. *A. oxydans* was reported as consisting of a specific A3α glycopeptide [19], the concentration of which might apparently correlate with the cell age.

E. coli is a common inhabitant of the lower part of the intestine of man and vertebrates. Many if not all strains of this gram-negative bacterium show opportunistic pathogenicity. In mature people, e.g., *E. coli* can be associated with peritonitis, gall bladder infections, inflammation of the pancreas, infections of the uro-genital tract, pneumonia and meningitis [3]. Figure 3 shows that *E. coli* grew much faster than *A. oxydans* in a MNB nutrient solution. However, similar to the former bacterium the composition of the FT-IR spectra became altered in course of culture aging (Fig. 4).

The intensity of C-H stretching vibrations in aliphatics at 3000–2800 cm^{-1} increased, similarly to that of NH_2 bendings and/or C-O, C-N stretchings at 1650 and 1540 cm^{-1} (amide I and II). A strong IR doublet at 1080–1030 cm^{-1} that was observed in the cell material from lag-phase could be assigned with carbonyl groups of cell wall glycopeptides and P-O-C group in phospholipids and esters. Already during the exponential phase of growth this

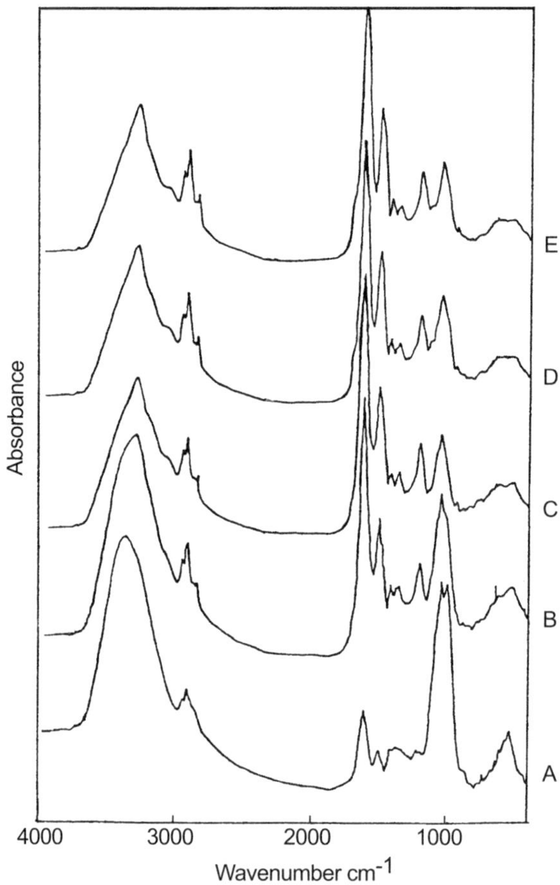

Figure 4. FT-IR spectra of *E. coli* cell mass harvested from MNB after 3 h (A); 16 h (B); 41 h (C); 72 h (D); 168 h (E).

band became much weaker. This feature indicates decreasing contents of some phospholipids as occurring during an early growth of *E. coli* [20]. At a same time a strong absorption of C-N stretching at 1235 cm^{-1} (amide III) at 1235 cm^{-1} has developed (Fig. 4, B-E).

An attempt was also made to estimate whether FT-IR characteristics would alter if bacteria are grown in differently composed cultural solution. *A. chroococcum*, a common free living soil bacterium capable of fixing atmospheric nitrogen, showed such effects (Fig. 5). If grown in a MNB nutrient solution enriched with glucose, *A. chroococcum* developed a well differentiated spectrum (Fig. 5, A) containing a strong triple-band at 1726 cm^{-1}, 1659 cm^{-1} (NH_2 bending, C-O in amide I), and 1537 cm^{-1} (C-N stretching, amide II). The absorption band at 1726 cm^{-1} could be attributed to poly-β-hydroxybutyrate, a polyester produced by *A. chroococcum* [10] rather than to carbonyl groups of long-chain fatty acid esters, for the absorptions at 2929 cm^{-1}, and C-H bending in fatty acids at 1380 cm^{-1} appeared only weakly.

Distinct bands of P-O stretching vibrations were detected at 1292 cm^{-1} and 1130 cm^{-1}. At 1235 cm^{-1}, C-N stretching in amide III absorbs. The IR absorption of *A. chroococcum* cell material was much weaker if the bacterium grew in a nutrient solution containing fructose instead of glucose (Fig. 5,C). If the culture medium was lacking a source of N, the

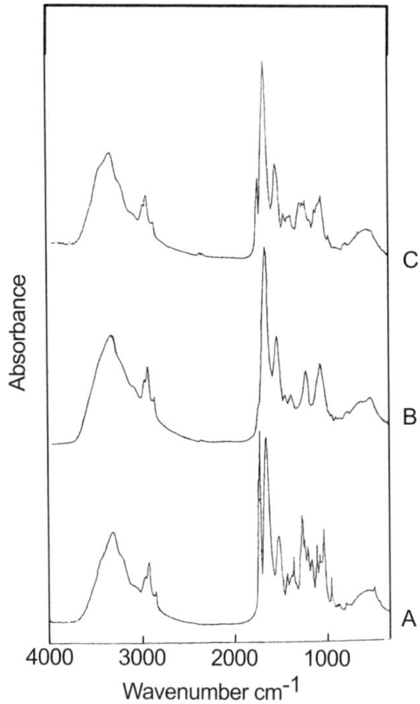

Figure 5. FT-IR spectra of *A. chroococcum* cultivated in MNB (A); in MNB without a source of N (B); in MNB with 40 mM of fructose instead of glucose (C).

structural composition of bacterial cells was altered again, as shown in Fig. 5, B. An absence of IR absorption at 1726 cm^{-1}, e.g., indicated that the formation of poly-β-hydroxybutyrate and perhaps some other cell constituents was suppressed. On the other hand, apparently due to the capacity of *A. chroococcum* to use molecular N_2 from the atmosphere, the IR bands of amide I, II and III remained sufficiently expressed in the respective spectrum.

We collected FT-IR spectra also from starved bacterial cells (see in Section 2.2) to eliminate possible influences of nutrient residues on the spectral patterns of bacterial biomass. Possible effects of the respective treatments should be demonstrated with the cell mass of *E. agglomerans*. In the cell mass of this bacterium, however, and similarly in other bacteria under testing, the starvation affected neither the composition of the IR spectrum, nor the intensity of the individual IR bands. Therefore we refrained from the demonstration of the spectra in a figure.

Conclusively to the topic of this paper, the FT-IR spectra of *A. oxydans, A. chroococcum, E. coli,* and *E. agglomerans* collected from the individual cultures grown in a MNB nutrient solution containing glucose are presented in Fig. 6. Simultaneously, wavenumbers, the assignments and relative intensity of IR absorptions as typical for the individual bacteria under testing have been listed in Table 1. The data show that the individual bacteria could be characterized by distinguished IR spectra in the waverange of 3400–500 cm^{-1}. The spectra of soil bacteria *A. oxydans* and *A. chroococcum* show distinct differences not only between each other but also relative to the spectra of *E. agglomerans* and *E. coli*. However, the spectroscopic differentiation failed between these two species of the *Enterobacteriaceae* family.

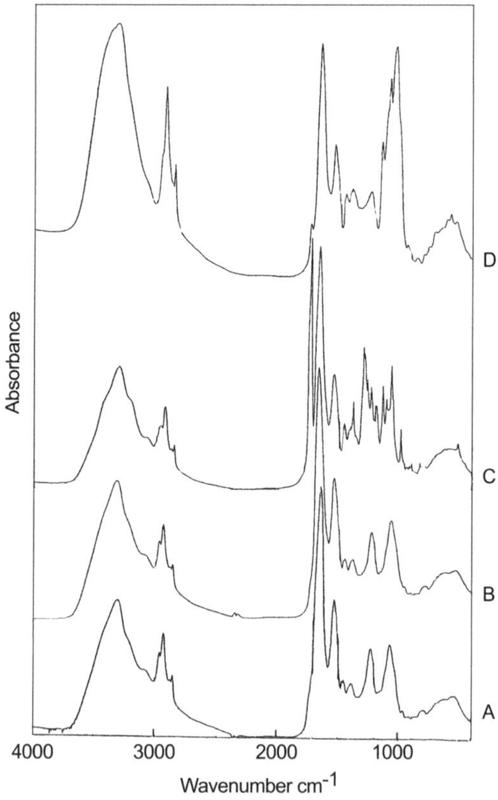

Figure 6. Standard FT-IR spectra of *E. coli* (A); *E. agglomerans* (B); *A. chroococcum* (C); *A. oxydans* (D) cultivated in MNB.

Conclusions

We were able to demonstrate that soil bacteria such as *A. chroococcum* and *A. oxydans*, and a two potentially pathogenic members of the family *Enterobacteriaceae* developed well differentiated FT-IR spectra, if uniformly grown in a Minimum Nutrient Broth. Therefore, the cultural conditions for enrichment of bacterial biomass should be duly standardized if attempting a differentiation between some soil and health relevant bacteria by FT-IR spectroscopy. Since the presence in soil of *A. chroococcum* is usually attributed to a good soil health and that of *E. coli* to a fecal contamination, the FT-IR spectroscopic differentiation of these bacteria could gain some interest in the detection of soil contamination.

Acknowledgements

The experiments were performed by S.H., and were supervised by the senior author (Z.F.) who wrote the manuscript in a co-operation with J.K. The senior author greatly acknowledges a Visiting Expert Award (EST.EV 980115) granted by courtesy of the NATO Public Diplomacy Division, Brussels, and a subsequent support from the Johann-Gottfried-Herder Foundation, Bonn, for his stay at the Mendel University of Agriculture and Forestry in Brno, Czech Republic.

Table 1. Infrared absorption bands of soil and health relevant bacteria under testing, their assignments and relative intensity (FT-IR spectra of cell mass harvested from a Minimum Nutrient Broth after 24 h).

Band (cm^{-1})	Possible assignment[1]	Intensity[2] of infrared bands in bacteria			
		A. oxydans	*A. chroococcum*	*E. agglomerans*	*E. coli*
980–800	H-bonded OH groups, NH_2 stretching (adenine, quanine, cytosine)	vs, br	m, br	m, br	m, br
2960–2850	C-H stretching in aliphatics (fatty acids)	vs, sh	w, sh	w, sh	w, sh
1720	-O stretching (saturated esters)	vw, sh	vs, sh	nd	nd
1660–1535	NH_2 bending, C-O, C-N stretching (amide I and II)	vs, sh	vs, sh	vs, sh	vs, sh
1467–1455	C-H deformations of CH_2 or CH_3 groups in aliphatics	vw, sh	v, sh	vw, sh	vw, sh
1402	C-O in carboxylate	vw, sh	nd	nd	nd
1390–1380	C-H bending, -CH_3 stretch (fatty acids)	m, sh	w, sh	w, sh	w, sh
1340	C-H bending (aliphatics)	w, sh	nd	nd	nd
1290	P-O stretching (phosphoesters)	nd	s, sh	nd	nd
1240–1230	C-N stretching (amide III)	m, sh	m, sh	s, sh	s, sh
1150–1000	C-O, PO^-_2 (glycopeptides, ribose)	vs, sh	m, sh	m, sh	m, sh
980–800	P-O-C, P-O-P stretching (phospholipids, ribose-phosphate chain, pyrophosphate)	nd	m, sh	vw, br	vw, br
550–515	C-O-C, P-O-C bonding (RNA, aromatics)	v, sh	v, sh	v, br	v, br

[1] Assignments according to references [18,22,23].

[2] vw = very weak, w = weak, m = medium, s = strong, vs = very strong, br = broad, sh = sharp, nd = not detected.

References

[1] J.R. Bicudo, S.M. Goyal, Pathogens and manure management systems: A review, *Environ. Technol.* 24 (2003) 115–130.

[2] E.N. Mishustin, M.I. Pertsovskaya, V.A. Gorbov, *Sanitary soil microbiology*, Nauka, Moscow, 1979 (in Russian).

[3] E. Mitscherlich, E.H. Marth, *Microbial survival* in *the environment*, Springer-Verlag, Berlin 1984.

[4] F. Möller, B. Stettnisch, K. Krannich, I. Härtel, Untersuchungen zur Überlebensdauer von Enterobakterien und Enteroviren in faulschlammgedüngten Böden, *Z. ges. Hyg.* 31 (1985) 237–241.

[5] K.P. Dierksen, G.W. Whittaker, G.M. Banowetz, M.D. Azevedo, A.C. Kennedy, J.J. Steiner, S.M. Griffith, High resolution characterization of biological communities by nucleic acid and fatty acid analyses, *Soil Biol. Biochem.* 34 (2002) 1853–1860.

[6] A. Ogram, X. Feng, Methods of soil microbial community analysis, in: Ch. J. Hurst (Ed.), *Manual of Environmental Microbiology*, ASM Press, Washington, D.C. 1997, pp. 422–430.

[7] Z. Filip, Infrared spectroscopy of two soils and their components, in : W.E. Krumbein (Ed.), *Environmental Biogeochemistry and Geomicrobiology*, Ann Arbor Science Publisher, Inc., Ann Arbor, Mich. 1978, pp. 747–754.

[8] Z. Filip, Infrarotspektren der mikrobiellen Biomasse und der Huminsäure im Podsolboden, *Z. Pflanzenernaehr. Bodenkd.* 141 (1978) 711–715.

[9] T. Hirschfeld, FTIR now and in the future, *Eur. Spectroscopy News* 51 (1983) 13–18.

[10] H.H. Manch, H.L. Casal, Biological applications of infrared spectrometry, *Fresenius Z. Anal. Chem.* 324 (1986) 655–661.

[11] D. Naumann, D. Helm, H. Labischinski, Microbiological characterizations by FT- IR spectroscopy, *Nature* 351 (1991) 81–82.

[12] P.D. Nichols, J.M. Henson, J.B. Guckert, D.E. Nivens, D.C. White, Fourier transform-infrared spectroscopic methods for microbial ecology: analysis of bacteria, bacteria-polymer mixtures and biofilms, *J. Microbiol. Methods* 6 (1985) 79–94.

[13] L.-P. Choo-Smith, K. Maquelin, T. Vreeswijk van, H.A. Bruining, G.J. Puppels, N.A. Ngo Thi, C. Kirschner, D. Naumann, D. Ami, A.M. Villa, F. Orsini, S.M.Doglia, H. Lamfarraj, G.D. Sokalingum, M. Manfait, P. Allouch, H.P. Endtz, Investigating microbial (micro)colony heterogeneity by vibrational spectroscopy, *Appl. Environ. Microbiol.* 67 (2001) 1461–1469.

[14] D.I. Ellis, D. Broadhurst, D.B. Kell, J.J. Rowland, R. Goodacre, Rapid and quantitative detection of the microbial spoilage of meat by Fourier transform infrared spectroscopy and machine learning, *Appl. Environ. Microbiol.* 68 (2002) 2822–2828.

[15] H. Oberreuter, J. Charzinski, S. Scherer, Intraspecific diversity of *Brevibacterium linens, Corynebacterium glutamicum* and *Rhodococcus erythropolis* based on partial 16S rDNA sequence analysis and Fourier-transform infrared (FT- IR spectroscopy, *Microbiology* 148 (2002) 1523–1532.

[16] H. Seiler, S. Scherer, FTIR-Spektrenbibliotheken für die Identifizierung von Mikroorganismen, *BIOspektrum* 9 (2003) 369–371.

[17] M. Wenning, H. Seiler, S. Scherer, Fourier-transform infrared microspectroscopy, a novel and rapid tool for identification of yeasts, *Appl. Environ. Microbiol.* 68 (2002) 4717–4721.

[18] F.S. Parker, *Infrared spectroscopy in biochemistry, biology and medicine*, Adam Hilger, London, 1971.

[19] H.L. Jensen, The Azotobacteriaceae, *Bact. Rev.*18 (1954) 165–209.

[20] C. Randle, P.W. Albro, J.C. Dittmer, The phosphogyceride composition of gram-negative bacteria and the changes in composition during growth. *Biochim. Biophys. Acta* 187 (1969) 214–220.

[21] K.H. Schleifer, O. Kandler, Peptidoglycan types of bacterial cell walls and their taxonomic implications, *Bact. Rev.* 36 (1972) 407–477.

[22] L.J. Bellamy, *Infrared spectra of complex molecules*, Chapman and Hall, London, 1975.

[23] R.M. Silverstein, G.C. Bassler, *Spectrometric identification of organic molecules*, Wiley and Sons, New York, 1964.

Emerging Biological Threat
G. Berencsi et al. (Eds.)
IOS Press, 2005

Advanced Medical Technologies Against Bioterrorism

M. GARSTANG[a] and E. BUSCH-PETERSEN[b]
[a]*Acambis plc, Peterhouse Technology Park, Cambridge, UK*
[b]*Baxter Vaccine AG, Industriestrasse 67, A-1220 Vienna, Austria*

Smallpox is caused by the Variola virus and is one of the most deadly diseases known to man, historically proving fatal in one out of three people who contract it. In the twentieth century alone, around 300 million people died of smallpox and many others were left disabled or scarred. The smallpox vaccine is based on *Variola*'s relative, the *Vaccinia* virus. First-generation smallpox vaccines were manufactured by crude methods from calf lymph.

In the 1960s eradicating smallpox as a disease became an international priority. In 1967 the World Health Organization (WHO) launched the Global Intensified Eradication Programme. The campaign was implemented over a decade and was ultimately successful with the last recorded natural occurrence of smallpox being in Somalia in 1977. On May 8, 1980, the WHO declared the global eradication of smallpox.

Once smallpox had been eradicated, the WHO decided to maintain a stockpile of smallpox vaccine, however, in 1986 it was decided that this was no longer needed. Bioterrorism was not an issue at the time, so the vaccine was returned to donor countries or destroyed.

Twenty years later, the threat of smallpox is back. The terrorist events of 9.11 led to serious concerns about the potential use of the smallpox virus as a bioterror weapon. Both Russian and American scientists had come to the conclusion that of all potential bioterror agents, smallpox posed the greatest threat. The threat of bioterrorism is now believed to be very real and the need to stockpile smallpox vaccine is increasingly seen as a necessity by governments around the world.

The characteristics of a potential bioterrorist attack from smallpox are considered to be that the attack is likely to be launched through self-infection. The likely prime targets would include localities with high population mobility. The location of impact may well remain unknown with no visible sign of the attack until the first and second wave cases have been correctly diagnosed, therefore, the agent may therefore spread undetected for several weeks.

The potential risk of smallpox has changed dramatically during the last forty years. Due to its eradication, few physicians today have experience in the diagnosis of smallpox. There is high population susceptibility due to the lack of natural exposure and vaccination during the past 30 years, i.e. no herd immunity. Furthermore a massive increase in commuting and international population mobility would significantly increase local and global spread.

In 2000, the Centers for Disease Control and Prevention (CDC) contacted the pharmaceutical company Acambis with an order to develop and manufacture 40 million doses of a new modern cell culture vaccine. This vaccine is currently in the investigational phase to evaluate it's safety and efficacy and will become part of the investigational stockpile. This contract called for continued warm base manufacturing for 20 years, so that the capability for surge capacity could be maintained. Following the events of 9.11 the contract was expanded to 54 million doses. The experience of Baxter in the field of continuous cell lines for the manufacture of vaccines placed Acambis and Baxter in a unique position to manufacture and supply smallpox vaccine for government stockpiling. In December 2001 a new contract for 155 million doses of investigational vaccine for the stockpile was agreed with Acambis/Baxter as the sole supplier to the U.S. Department of Health and Human Services. This product is currently in the investigational phase of clinical trials

The objectives as defined by the U.S. government for the stockpiling of smallpox vaccine are to ensure optimal protection of the total population in the case of an emergency and to avoid panic by being able to give a reassuring message of one dose per citizen.

Other considerations for the U.S. government, in addition to the timely manufacture of vaccine, were national logistic issues in getting the vaccine to the points of vaccination, the training of healthcare workers and public information/training.

The new Acambis/Baxter investigational vaccine, ACAM2000, is a second-generation vaccine grown on continuous vero cell lines, in a serum-free cell culture. ACAM2000 is a lyophilised preparation of live, attenuated vaccinia virus. The ACAM2000 strain of vaccinia virus was derived by plaque purification cloning from Dryvax® (Wyeth Laboratories, Marietta, PA, calf lymph vaccine, New York City Board of Health Strain)

ACAM2000 is reconstituted by addition of 0.3mL of sterile glycerol-phenol diluent. The diluent, supplied in clear glass vials, is composed of 50% (v/v) Glycerin USP and 0.21% (v/v) Phenol USP in Water for Injection. The vaccine and accompanying diluent should be removed from cold storage and brought to room temperature before reconstitution of the vaccine. The flip cap seal of the vaccine and diluent vials should be removed, and the rubber caps are wiped with an isopropyl alcohol swab and allowed to dry thoroughly. Using aseptic technique and a 1 mL syringe fitted with 25 gauge x 5/8" needle is used to draw up 0.3 mL of Smallpox Vaccine Diluent and the entire contents of the syringe should be transferred to the vaccine vial. The mixture should be gently swirled to mix but effort made not to get the product on the rubber cap. After reconstitution, each vial of ACAM2000 solution contains 100 nominal doses. The concentration of vaccinia virus is $1.0–5.0 \times 10^8$ plaque-forming units (PFU)/mL. The reconstituted solution is a straw-coloured, clear liquid, pH 6.5–7.5.

ACAM2000 is administered by the percutaneous route. The vaccine (approximately 0.0025 mL held between the tines of a bifurcated needle) is deposited on the surface of the skin over the deltoid region and inoculated by scarification of the epidermal layer of the skin, using 15 strokes of a sterile bifurcated needle.

ACAM2000 should be stored centrally at between -10° C and -25° C. For distribution and use in the field, the vaccine may be transported and stored at +2 to $+8^o$ C. After reconstitution, ACAM2000 vaccine may be stored in a refrigerator and used for up to 30 days, after which it should be discarded.

A comprehensive clinical trial programme is being conducted for ACAM2000. Phase I and II trials compared ACAM2000 with Dryvax to evaluate safety, tolerability and immunological response in subjects both with and without previous smallpox vaccination. A phase III programme was initiated in 2003The potential use of non-licensed IND vaccines requires special attention to liability and indemnification issues. Vaccines purchased for stockpiling against bioterrorist agents will commonly be unlicensed, such as ACAM2000, are intended for emergency use, must be fully covered by the purchaser with regard to liability and indemnification and should be targeted for full clinical testing aimed at licensure.

Emerging Biological Threat
G. Berencsi et al. (Eds.)
IOS Press, 2005

Science for Peace Means International Cooperation to Control and Prevent Emerging Natural or Bio-Terror Virus Epidemics with Attention to the HIV-1/AIDS Pandemic

Professor Yechiel BECKER
Director, UNESCO-HUJ International School of Molecular Biology and Microbiology
The Institute of Microbiology, the Department of Molecular Virology,
Faculty of Medicine, The Hebrew University of Jerusalem,
POB 12272, Jerusalem 91120, Israel
Tel: 972-2-6758394 Fax: 972-2-6784010
Email: becker@md.huji.ac.il

The control of natural and bio-weapon-related virus epidemics require international governmental treaties, scientists who are dedicated to science for peace, and national health services cooperating with the World Health Organization (WHO). Unfortunately, the signatories of the Biological Weapon Convention (BWC) that prohibits the production and stockpiling of bioweapons has failed to reach an agreement on the need to inspect and control the related industries. As a result, fears of terrorist attacks involving bioweapons are mounting. In addition, the UNESCO Science-Agenda-Framework for action, which puts the responsibility to protect against misuse of biological knowledge on scientists and students of science has yet to be implemented.

On the other hand, international collaborations with WHO on the control and eradication of smallpox, poliomyelitis, measles, influenza, and recently SARS have been successful. Yet, the marked increase in the number of allergic and immune deficient people in industrial countries restricts the use of smallpox vaccine (vaccinia virus) for protection of unimmunized populations against a possible terror attack with a smallpox bioweapon. It is indicated that international cooperation with WHO in the event of a smallpox bioterror attack is needed to localize the infection and prevent it from spreading in the population.

The HIV-1/AIDS pandemic currently requires novel approaches to understand why infected people are not able to control the virus infection and recover. There is an urgent need to develop new approaches to stop the HIV-1/AIDS pandemic and save infected people.

1. International Efforts to Prevent the Misuse of Biological Sciences

a. Conceptual approaches to providing guidelines against bio-weapons.
(i) *The concepts in the Jerusalem Statement on Science for Peace (1997).*

Members of the international scientific community met in Jerusalem, Israel on 20–23 of January 1997 to participate in the Second International Symposium on "Science for Peace." The Symposium was organized by the UNESCO Hebrew University of Jerusalem (HUJ) International School for Molecular Biology and Microbiology (ISMBM), directed by Professor Yechiel Becker, with additional support of UNESCO Paris, UNESCO Venice Office, the UNESCO Global Network for Molecular and Cellular Biology, the International Institute of Theoretical and Applied Physics (IITAP) at Iowa State University, and the Hebrew University of Jerusalem [1].

The members of the Symposium on Science for Peace concluded the deliberations by issuing the Jerusalem Statement on Science for Peace (1997) that is addressed to all individuals and institutions working in and for science. As the language of science is universal and cooperation in science builds important bridges of communication, we appeal for increased and unified efforts to adopt Science for Peace as an important principle. The recommendations are as follows:

1. Scientific endeavors and achievements will only be used for peaceful purposes and for the greater benefit of humanity.
2. Members of the academic community will be afforded the freedom to travel anywhere they so choose.
3. The free flow and sharing of scientific information and knowledge.
4. The academic environment remains open and dedicated to the free expression of ideas.
5. Efforts are to be undertaken to develop a "Science for Peace Oath" that young scientists are to take upon accepting their degrees [1].

The Jerusalem Statement on Science for Peace (1997) was approved by the director general of UNESCO, Professor Federico Mayor and the assistant director general for science, Maurizio Iaacarino, and was presented to the UNESCO committee that prepared the UNESCO declaration on Science and the Science Agenda-Framework for Action that concluded the UNESCO World Conference on Science for the Twenty-first Century, a New Commitment, Budapest, Hungary, 26 June–1 July 1999.

(ii) *The UNESCO Science Agenda-Framework for Action.*

The concepts on Science for Peace in the Jerusalem Statement on Science for Peace were accepted and the committee further expanded the concepts on Science for peace and Development (Section 2 of the agenda). Paragraph 2.5 entitled "Science for peace and conflict resolution," states the following:

1. "The basic principles of peace and coexistence should be part of education at all levels. Science students should also be made aware of their specific responsibility not to apply scientific knowledge and skills to activities, which threaten peace and security" (item 50).
2. "Military and civil sectors, including scientists and engineers, should collaborate in seeking solutions to problems caused by accumulated weapon stocks and landmines" (item 53).

In paragraph 2.1, "Science for basic human need," it is stated:

4. All countries should share scientific knowledge and cooperate to reduce avoidable ill-health throughout the world" (item 26) [2].

These are a few examples dealing with the concepts of Science for Peace, which are the guidelines to national government and to scientists and engineers.

Unfortunately, the committee members did not include the Science for Peace Oath in the Framework for Action, although they had indicated that Science for Peace concepts should be part of education at all levels and science students should not apply scientific knowledge and skills to activities which are against peace and security. What is needed to achieve these goals is to reach an international agreement that the development of educational approaches for the teaching of Science for Peace concepts and an international agreement to demand that upon graduation all students will take the Science for Peace Oath.

b. International cooperation in preventing emerging and re-emerging microbial diseases.

(i) *The UN World Health Organization (WHO) and their role in monitoring and preventing emerging diseases.*

The importance of WHO surveillance for pathogenic influenza virus isolation by a central WHO influenza virus laboratory that collaborates with national diagnostic laboratories around the world in monitoring influenza virus infections is important in the prevention of an influenza epidemic/pandemic. Another function of WHO is the ability to dispatch expert teams to countries where the Ebola virus outbreaks have infected many people and treatment of patients with antisera against Ebola virus to stop the spread of the virus. The example of WHO initiative in India and the Horn of Africa to immunize the entire populations in which smallpox disease was endemic led to the eradication of smallpox in 1977. The WHO announcement that smallpox disease was eradicated relieved the worldwide human population from the need to use vaccinia virus to vaccinate the entire population against smallpox [3].

Due to the WHO decision that prevailed until the terror attack on 11 September 2001 on the United States and around the world, and the large number of people that are unimmunized against smallpox increased the fear that terrorist organizations will use smallpox virus (variola major) as a bioweapon. This prospective led to the decision to immunize the USA population with a smallpox vaccine [4]. The decision was taken despite the available information that people with immune deficiency diseases should not receive the vaccine. During the session on smallpox vaccination at the 2002 International Congress Virology in Paris, I commented that in the case of bioterror with smallpox any place throughout the world, it is the responsibility of WHO to dispatch a special intervention team with specific antibodies and vaccine virus to treat smallpox patients and to help the national health authority to control the spread of the disease. I received an e-mail from Dr Diego Buriot, director of the WHO CSR office in Lyon, France, Department of Communicable Disease, Surveillance and Response (CSR) that their mission is to help national health authorities control a terrorist smallpox attack.

Unfortuantely, at the onset of the HIV-1/AIDS epidemic, WHO was not able to control the epidemic due to the absence of knowledge about the virus and the disease. Now, HIV-1 virus is changing constantly and is causing a worldwide pandemic that is predicted that seventy million infected people will die by 2020.

(ii) *The United Nations Biological Weapon Convention (BWC).*

Parallel to the development of the international Atomic Weapons Convention and the Chemical Weapons Convention (BWC) on 10 April 1972, the Biological Weapons Convention was open for signatures from the UN membership [5]. The BWC was signed by many UN member states. However, the Soviet Union and Iraq continued to develop biological weapons as was found after the decline of the Soviet Union [6] and after the 1991 Desert Storm War against Iraq. Further deliberations of the BWC committee members on the need for inspectors to be able to monitor the industries that produce biological products similar to the inspections of atomic and chemical facilities were held. Unfortunately, an agreement to allow inspectors free access to national biological industries has failed. The acts of terror against the United States on 11 September 2001 and the spread of anthrax through American mail became a grim reminder for the need of a strong BWC convention that can be enforced and controlled.

(iii) *The need for an international commitment from United Nations member states to prevent the misuse of biological sciences.*

The Jerusalem Statement on Science for Peace [1] proposed the concept of the "Science for Peace Oath" for scientists and the UNESCO Science Agenda-Framework for Action [2] had stated that science students should be aware of their specific responsibility not to apply scientific knowledge and skills that threaten peace and security. It also emphasized that, "The basic principles of peace should be part of education at all levels" [2]. Without the development of an educational system for Science for Peace and without a Science for Peace Oath, the UNESCO "Framework for Action" cannot be fulfilled. International efforts are needed by UNESCO, NGOs and universities to introduce the concepts of Science for Peace and the responsibilities of the scientists at all educational levels, especially at universities worldwide.

As we have noted from the past activities of UN organizations, WHO is active and well connected to national health systems worldwide in distributing and receiving information on problems of emerging and reemerging pathogens. The development of the BWC, an arm of the UN Security Council that failed to act on the subject of international control of bioweapons is urgently needed. UNESCO has yet to start developing an educational system for all levels of the educational system to introduce the concepts of Science for Peace, especially at the university level.

In view of the possibility of the use of smallpox as a bioweapon by terrorists, the present study will analyze what was learned from the attempts to control the recent SARS epidemic without antiviral drugs or vaccines; from the control of smallpox pandemic almost two-hundred years after Edward Jenner presented the smallpox vaccine in 1896; and the unsuccessful attempts to develop vaccines to control the spread of HIV-1/AIDS during the last twenty years.

2. What Has Been Learned from Current and Past Epidemics/Pandemics to Prevent Bioterrorism-Related Virus Epidemics?

5. The SARS epidemic started in South China and was carried by travelers to nearby and distant countries.

(i) *Mutant influenza viruses from fowl infect human beings.*

During the twentieth century, respiratory virus infections in fowl and humans were prevalent in southern China. The constant threat was the re-assortment of influenza virus in ducks and chickens that harbor and maintain many influenza virus mutants with the potential to infect humans. The past events involving the emergence of highly pathogenic virus strains, which caused pandemics during 1957 and 1968, require constant monitoring of the influenza virus mutants in ducks, geese, chicken, and wild fowl in China. An outbreak of chicken influenza in eighteen people was noted during 1997, which killed six of the infected individuals. To stop the epidemic before it spread, the Honk Kong authorities killed some one million birds, and the pathogenic virus was prevented from spreading to human populations. The influenza virus strain that caused the human infection was identified as influenza strain H5N1 (Heremagglutinin #5 and neuraminidase #1). A recent avian influenza virus outbreak occurred on a Dutch poultry farm and the isolated virus was designated as H7N7 that was thought to be with a low transmission efficiency to humans. However, since on 28 February 2003, more than eighty poultry workers were infected and developed conjunctivitis, and on 17 April a veterinarian who had visited poultry farm had died of pneumonia. To prevent the spread of the H7N7 influenza virus, more than twenty million fowl were slaughtered [7]. When the first human cases were found in southern China, it was assumed that they were infected with influenza virus, but the clinical symptoms were different.

(ii) *A new syndrome of "severe acute respiratory syndrome (SARS)" in humans in China, 2003.*

The new epidemic started in China with four individuals that were hospitalized in Fusan (16 November 2002), Shunda (2 December 2002), Gunszu (3 December 2002), and Zongshan (2 January 2003). The patient form Zongshan was treated in the emergency rooms of several hospitals (2–4 February 2003) and infected tens of health workers, before he was subsequently hospitalized in one of the hospitals. The persons form Shunda was a chicken and snake dealer who infected his wife and medical personnel at the hospital. The patient in Zongshan was a cook who infected thirteen members of a medical team [8]. These patients were treated by hospital teams that did not use protective gear, despite the fact that it is common knowledge that respiratory viruses are easily transmitted from person to person.

A physician, Dr. Liu Jiang-Lon a lung expert at Zongshan hospital traveled to the Hotel Metrople in Hong Kong, where he developed the SARS syndrome and infected other guests. Three guests traveled to Singapore at the end of February 2003 and were subsequently hospitalized in three separate hospitals. One guest, Ms Esther Mack, infected over ninety people.

An American businessman, Mr. Johnny Chen, another guest at the Hotel Metropole, traveled to Hanoi, Vietnam where he developed the SARS disease and infected twenty health workers over the course of his hospitalization. Furthermore, another guest, Ms. Quan Soy-Chu, traveled to Toronto, Canada where she was stricken with SARS. She subsequently infected her son and five health care workers during her hospitalization before dying.

At this point, the unknown virus that caused the new acute respiratory syndrome had spread through China, Hong Kong, Vietnam, Singapore, and Canada and threatened the rest of the world [8]. By late April 2003, the WHO received over 4,300 reports of SARS cases as well as 250 SARS-related deaths from over twenty-five countries around the world. The incubation period was two to seven days and death form progressive respiratory failure occurred in 3% to 10% of the patients [9]. WHO coordinated an international collaboration, including clinical, epidemiological laboratory investigations, and efforts to control the spread of the SARS.

The identification of the SARS causative agent was successful during March 2003 due to the collaborative research efforts in the United States, Canada, Germany, and Hong Kong. A novel coronavirus (SARS-COV) was isolated from SARS patients and their sera contained antibodies to the virus [11]. The virus genome was sequenced and the open reading frames were identified [9,10], but the viral genes which code for proteins that are responsible for virus pathogenicity were not yet identified [12].

The SARS-COV virus was discovered and isolated from Chinese masked palm civets that were on sale in the Guandong market, possibly the natural host [13].

(iii) *How the SARS epidemics were controlled in the absence of vaccine and/or antiviral drugs?*

Dye and Gay [14] modeled the public health measures that were taken to protect health workers treating SARS patients, which included the wearing of facemasks and gloves, disinfecting the waste material, and the cleaning of buildings and public areas. The public was also ordered to wear masks. By late March 2003, about four months after the first index SARS case was hospitalized, the SARS-Corona Virus was isolated, identified and analyzed. It was found that the virus particles are coated with a membrane that is sensitive to detergents. Therefore, the use of detergents to treat patients' excretions and clothes might have inactivated the virions and prevented the virus from spreading in the human population. The cleaning buildings and the environment in Hong Kong and other places brought the SARS epidemic to an end. The decline in the number of air commuters and the use of masks by those who boarded intercontinental flights also helped prevent virus spreading.

The control of the SARS CO-virus needs further surveillance for re-emerging SARS infections. There is a need to develop effective antiviral drugs [15] and a vaccine to immunize people in endemic areas against re-emerging SARS infections.

b. The smallpox pandemics (1160 B.C.–1977 A.D.):

(i) *From Jenner's discovery of preventive smallpox vaccine (1798) until the smallpox pandemic was eradicated (1977) 220 years elapsed.*

The long-term epidemic that can be dated back to 1000 B.C. is the smallpox pandemic that left characteristic pockmarks on the face of the mummified head of Egyptian Pharaoh Ramses V [16]. Elgood [17] wrote that the Greek physician Gallen in the second century A.D. and the Hebrew physician El Yehudi, who lived during the seven century A.D., described in their writing smallpox epidemics. The Persian physicians, Muhamad ibn Zakariyya (Rhazes) who was born in 865 A.D. and lived in Baghdad, described smallpox as an epidemic that was brought to Arabia from Abysinia.

During the sixteenth century, smallpox disease spread in England and during the eighteenth century, smallpox was spread in many parts of the world. Smallpox virus (variola major) was introduced to Central American populations during the European conquest by the Spanish soldiers who brought the disease from Europe. Between 1519–1520, five to eight million people perished in Mexico. A second catastrophic epidemic from 1545–1576 killed an additional seven to seventeen million people. The epidemic with a high mortality was estimated to have killed 5–15 million people (80% of the native Mexican population) more than those who perished during the "Black Death," (bubonic plague) from 1347–1351 in western Europe, which killed approximately twenty-five million people (50% of the regional population). However, the 1576 epidemic that killed over two million people in Mexico is now thought to have been caused by hemorrhagic fevers [18].

The method of "variolation," the use of infectious material from pocks on the skin of smallpox-infected patients who survived, for inoculation into the skin of uninfected individuals was known in China and the Middle East, but not in Europe. During the smallpox epidemic in the second half of the eighteenth century, the country doctor Edward Jenner noted that milkmaids were not infected by the smallpox. On May 14, 1796, Jenner inoculated the arm of eight-year old James Phipps, with material obtained from a cowpox lesion on the hand of Sara Nelmes, a Gloucestershire milkmaid. Six weeks later, Jenner inoculated the boy Phipps with material from a smallpox victim and found that Phipps was resistant to the smallpox infection. In 1798, Jenner published his findings and suggested that smallpox disease was caused by a "virus" (poison in Greek). Many years later, Jenner's smallpox vaccine became the principal prophylactic vaccine for the protection of humans against smallpox [19]. In thirty-three countries, mainly in Asia and Africa, smallpox remained endemic. The WHO program for smallpox disease eradication by immunizing every person living in smallpox endemic areas led to the eradication of the disease, and the last smallpox-infected child was found in 1977 in Somalia. The "victory was brought around by hundreds of thousands of health workers all over the world, eradication is a triumph of international cooperation and preventive medicine," wrote D.A. Henderson [20].

(ii) *Hopes after the victory over smallpox were shattered by the reality of early twenty-first century fear of bio-terrorism.*

Halfdan Mahler, the director-general of WHO during the last smallpox eradication campaign wrote twenty-three years ago in his 1980 article [21] entitled, "Towards the year 2000" that, **"Victory over smallpox has implications that go far beyond the individuals directly concerned, however. It reasserts our ability to change the world around us for the better, through mutual collaboration and mobilization of resources, allied to human energies and the will to succeed. It comes like a freshening wind for a vessel too long becalmed, creating new impetus as we set our course toward health for all by the year 2000."**

The happiness generated by smallpox eradication quickly changed within a year due to the appearance of patients in Africa, Europe, and the United States who suffered from a new disease syndrome that was named acquired immune deficiency syndrome (AIDS) (see below). What the director general of WHO could not have known, as the evidence was only provided in 1999 by Ken Alibeck, was that the response of the bioweapon developers in the former Soviet Union was of a negative kind. Alibeck wrote in his book "Biohazard" [22]: "The conquest of smallpox generated a special feeling of accomplishment in the Soviet Union: the worldwide crusade against smallpox has been a Soviet initiative. Moscow first proposed the campaign at a WHO meeting in 1958… Soon after the WHO's announcement, smallpox was included in a list of viral and bacterial weapons targeted for improvement in the 1981–1985 Five-Year Plan. Where other governments saw medical victory, the Kremlin perceived a military opportunity. A world no longer protected from smallpox was a world newly vulnerable to the disease" [22].

The stockpiling of bioweapons by the Soviet Union and the bioweapons developmental attempts by Saddam Hussein's regime in Iraq (discovered during the 1991 Desert Storm War) increased the anxiety of the Western world fearing a possible bioweapons attacks by terrorist nations and organizations. Twenty years after eradication of smallpox, millions of children and young adults between the ages of two and twenty-five are un-immunized and vulnerable to smallpox infection. The need to vaccinate this group as well as the rest of the population with smallpox vaccine (vaccinia virus) against smallpox bio-terror is clear. However, such a program is costly and problematic.

(iii) *The problems with the smallpox vaccines and human vaccination.*

Mair et al. [23] reviewed the existing stocks of smallpox vaccine — fifteen million doses of Dryvax® (Wyeth Labs) and eighty-five million doses of Glycerated Vaccine (Avenis-Pasteur) — that can be diluted to provide sufficient doses to immunize the entire United States population. However, Dryvax® is the only vaccine currently licensed by the FDA and the licensure does not yet include diluting the vaccine beyond 1:1. Nevertheless, the United States government has ordered 209 million doses of new vaccine, ACAM 1000 and ACAM 2000 (Acambis-Baxter), but FDA licensure is not expected until 2004. Even with the cessation of bioweapon production in Russia and the downfall of Sadam Hussein's regime in Iraq, the Al-Qaida terror organization may still use bioweapons. Moreover, the protection of the population of one country does not prevent smallpox-infected people form traveling to other continents and spreading the epidemic.

The downfall of Saddam Hussein's regime in Iraq is a problem since the stocks of the Iraqi bioweapons have yet to be uncovered since terrorist organization may get hold of the stockpiles of bioweapons. The need will arise to use smallpox vaccine in the event of an epidemic. We need to remember that the vaccinia virus is a live virus that has been passaged in calf skin, embryonated chickens, eggs, or tissue cultures during the twentieth century to produce large vaccine stocks. In addition, the molecular and immunological investigations on variola major virus (the cause of smallpox) and on vaccinia virus (vaccine) revealed that these two viruses resemble each other by having numerous genes, which the viruses use to evade the human protective innate and adaptive immune systems [24]. The major difficulty with smallpox vaccination is the marked increase in the number of immune deficient people and their household members who are not allowed to be vaccinated. Thus, the availability of smallpox vaccine does not mean that an entire national population can be vaccinated. Lastly, vaccinia virus originated from pocks on cattle and horses, and therefore vaccinated people may pass on the vaccinia virus to domestic and wild animals. Recently, it was reported from Brazil that vaccinia-like poxviruses infected people who take care of cattle. The virus was identified to be vaccinia virus strain WR, the vaccinia virus that was used for human vaccination in Brazil during the smallpox eradication program [25].

(iv) *Why does the vaccinia virus vaccine cause complications when it is inoculated to allergic and immune deficient persons?*

a. People who should not get the smallpox vaccine.

Engler et al. [26] wrote that due to the threat of bioterrorism with pathogenic microbes, such as smallpox virus (variola major), the question of the importance of widespread voluntary vaccination with smallpox vaccine needs to be asked. A major challenge lies in the ability to protect the population from disease, while minimizing the considerable side effects from the vaccine. Individuals with active or quiescent atopic dermatitis are at increased risk of vaccinia complications like the life threatening eczema vaccination.

Based on the experience drawn from past vaccination programs, the Center of Disease Control (CDC) released a "Smallpox Fact Sheet" listing people with medical conditions who should NOT get smallpox vaccines (<u>unless</u> they are exposed to smallpox):

6. Eczema or atopic dermatitis, even when the condition is not active, mild, or during childhood.
7. Skin conditions such as burns, chickenpox, shingles, herpes, severe acne, or psoriasis.
8. Weakened immune system (cancer treatment, organ transplant).
9. HIV/AIDS, primary immune deficiency disorders and medications to treat autoimmune disorders.
10. Pregnancy or plans to become pregnant.

Mair et al. [23] reported that as many as 10,000,000 individuals in the United States may be immunocompromised and at risk of sever complications from smallpox vaccination, including approximately 8,500,000 cancer patients, 850,000 individuals living with HIV infections or AIDS, 184,000 solid organ transplant recipients, and all persons who suffer from atopic dermatitis, eczema, and allergies. It has been estimated that these individuals and their household contacts were excluded from vaccination, approximately 50% of the United States population would be excluded from a preemptive, voluntary program.

(v) *Why vaccinia virus causes a mild skin infection in individuals with a normal immune system and in atopic dermatitis patients a life threatening complication?*

In a letter to ASM News, Bernard Moss [27] wrote that the "Number of immune-evasion proteins encoded by vaccinia virus and variola virus are similar, whereas cowpox virus encodes many more, but nevertheless produces a limited infection in humans." This is based on the assumption that poxivirus pathogenicity is a viral function. However, since vaccinia virus infection in the human body depends on the activities of the host innate, cellular, and humoral immune responses, a decline of one or all immune responses by providing a milieu that may enhance virus pathogenicity.

The answer to the increased vaccinia virus pathogenicity to patients with atopic dermatitis was deduced from the studies on the effect of interleukin-4 (IL-4) on the pathogenicity of vaccinia virus in mice. Sharma et al. [28] reported that expression of the mouse IL-4 gene (that was cloned into the genome of vaccinia virus) by the recombinant virus in infected mice inhibited cytokine synthesis T helper 1 cells, and as a result the development of the antiviral cytotoxic T cells (CTLs) response is prevented. The anti-vaccinia virus CTLs are responsible in the normal mouse for the clearance of the virus from the virus infected skin and the prevention of the virus spreading in the mouse. However, in the absence of antiviral CTL activity in atopic patients, vaccinia virus continues its invasion of essential organs and the disease progression causes a life threatening disease.

Becker [29] reviewed the studies on the immune status of atopic dermatitis patients. Numerous studies reported that environmental allergens activate T helper 2 cells to synthesize the cytokines IL-4, IL-5, and IL-13. The increase of IL-4 content in atopic dermatitis patients and other allergies is responsible for the inhibition of the anti-vaccinia virus CTL response and the enhancement of vaccinia virus pathogenicity. It is imperative to protect allergic people from vaccinia virus.

(vi) *What should be done to protect immune deficient people from the pathogenicity of the vaccinia virus smallpox vaccine?*

Allergens induce in sensitive people the activation of Th2 cells to synthesize the cytokines IL-4, IL-5, and IL-13. IL-4 induces B cells to synthesize anti-allergen IgE and the induction of IL-4 responsive genes. Thus, it is logical to assume that treatment of allergies with an immune response modifier that is capable of inhibiting the synthesis of T helper 2 and activating Th1 cells may be beneficial to the patients. The reversion of the IL-4 induced inhibition of Th1 will allow the patients to resurrect the antiviral CTL response to clear the vaccinia virus infection.

The reality of the prevalence of immune deficient people has led the United States Center of Disease Control (CDC) to update its vaccination plans and guidelines. The vaccination plans are to use a "ring vaccination" after identification of a smallpox-infected patient. When patients are confirmed to be infected with the smallpox virus, contacts are traced, vaccinated, and kept under close surveillance. Moreover, appropriate measures such as local quarantine and travel restrictions may be enforced. The plan does not recommend mass vaccination campaigns [30].

Let us hope that smallpox bioterrorism will not materialize, since bioweapons have never been used against civilians and armies during the twentieth century.

3. How to Stop the HIV-1/AIDS Pandemic that Started in 1980?

a. A short history of HIV-1 and HIV-2 in Africa.

(i) *Origins of HIV-1 and HIV-2.*

In a 1980 review, Essex and Kauki [31] reported that simian immune-deficiency virus (SIV) was isolated from monkeys and was found to be 50% homologous to HIV nucleotide sequences. Eleven years later, Gao et al. [32] reported that SIVcpz viruses related to HIV-1 were isolated from three living chimpanzees Pan t. troglodytes in central and eastern Africa. It was also found that HIV-1 clade N is a mosaic of SIVcpz and HIV-1 nucleotide sequences, indicating ancestral recombination event in the chimpanzee host [32] or possibly in the human host by integration of SIVcpz into a human endogenous retrovirus in the genome of a SIV-infected person.

The origin of HIV-2 in West Africa was reported by Marx et al. [33] who isolated the SIVsm strain LIB1 from pet sooty mangabeys. The isolated virus was found to be a member of the SIVsm/HIV-2/SIVmac group. The HIV endemic area in West Africa is a co-location of the P. Sooty Mangabeys in West Africa. The lack of disease in the monkeys suggested cross species transmission.

These studies revealed that the highly pathogenic HIV-1 may have infected local people who handled chimpanzee organs for food, and once the virus recombinant adapted itself to the human host the transmission of the virus in the human population by sexual infections has started. As long as the HIV-1 infected people lived and died in remote jungle settlements, the disease was not recognized although European people who traveled between Africa and Europe developed a disease and died in Europe of an unknown infection during the 1950s.

(ii) *The spread of HIV-1 and the development of a Mega-pandemic.*

The contact between American people with HIV-1 infected Africans led to the transfer of HIV-1 to the people in San Francisco in whom an unrecognized disease syndrome became apparent. In Europe, at that time an unknown disease was noted only in hospitalized patients from East Africa who were living in Paris and Brussels. From such patients, the Pasteur Institute team in 1983 had isolated the HIV-1 virus [34].

Between the discovery of HIV-1 in the early 1980s to the development of diagnostic tests for HIV-1 a few years later, HIV-1 infections spread in Africa due to the sexual transmission by infected truck drivers traveling between the east and west coasts. In the United States, drug users were infected. Moreover the blood banks obtained HIV-1 contaminated blood samples, which also enlarged the HIV-1 infected population.

Today, the HIV-1 pandemic can be regarded as a mega-pandemic. The WHO and the Joint United Nations Programme on HIV/AIDS (UAIDS) estimated that if the pandemic proceeds at its current rate there will be forty-five million new infections by 2010 and nearly seventy million deaths by 2020 [35].

Schwarländer et al. [36] reported on the AIDS situation in the world at the end of 1999, showing that 34.3 million adults and children are living with HIV-1/AIDS. In fifty-two countries, more than 1% of all adults carry the virus. The spread of HIV is most severe in Sub-Saharan Africa, where the AIDS epidemic started and recently the virus has spread to the Horn of Africa.

(iii) *Development of HIV-1 recombinants and drug resistant strains that drive the expanding mega-epidemic.*

Infection of the same T cell with more than one HIV-1 strain generates new recombinant viruses with mosaic nucleotide sequences of two or more HIV-1 strains from different clades [37]. The HIV-1 recombinants are different forms of the wild type HIV-1 strains and may have different effects on the immune system of the infected individual.

The expanding treatment of HIV-1 infected people with antiviral drugs helps prolong the lifespan of infected people, but also select drug resistant mutants that may reduce the efficacy of these drug treatments.

b. The HIV-1 enigma.

(i) *HIV-1 infection in humans and SIV in monkey are regular lentivirus infections to which the hosts respond by eliciting antiviral cytotoxic T cell responses a few days after infection.*

The unsolved riddle of HIV-1 is the question of the inability of the cellular immune response to clear the virus from the infected lymph nodes and the blood although effective antiviral CTLs are present in the host. Similarly, the human anti-viral humoral response to the infecting virus is not effective, and the HIV-1 virus continues to replicate almost undisturbed in the host.

What is not fully accounted for is the synthesis of IgE molecules in an increasing rate prior to the development of AIDS as well as the decline of the innate immune, cellular, and humoral immune responses.

(ii) *Why the attempts to develop protective and curing HIV-1 vaccines have failed?*

Albert Sabin [38] in his last published paper wrote: “The main challenge is to find a way to kill cells with chromosomally integrated HIV cDNA without harming normal cells.”

Chinen and Shearer [39] reviewed the strategy of vaccine development of nine vaccine candidates: four vaccines contained HIV envelope glycoproteins; three contained viral Gag or pol proteins (with or without Gag protein); and two vaccines contained SIV envelope proteins for the immunization of monkeys. The viral antigens were either recombinant peptides, DNA vaccines coding for viral proteins, or virus recombinants as expression vectors containing HIV-1 genes, Venezuelan equine encephalitis virus (VEE), poliovirus, vaccinia virus or canarypox viruses.

Cohen [40] reported on the vaccine trial by the California-based VaxGen Company’s five thousand-person study that conclusively showed that the product failed to protect the vaccinees against HIV infection. The unsuccessful vaccine trial led to the decision of the HIV vaccine trial network not to continue with the plans for the testing of the canarypox-HIV vaccine trial that was produced by the Avantis-Pasteur company [41].

An attempt to develop a live attenuated vaccine with multiply deleted HIV genes was reported by Baba et al. [42] to cause AIDS in infant and adult macaques. The authors concluded that this multiply deleted SIV is pathogenic and that human AIDS vaccines built of similar prototype virus may cause AIDS.

It seems logical that the continuation of the development of HIV-1 vaccines requires the elucidation of the reasons for the inability of the viral proteins to function as effective immunogens. There are two possibilities: 1) the viral proteins damage the host humoral immune system; or 2) the viral proteins contain immunogenic domains, but the host immune system responded to the viral antigens in a way that prevents the development of effective anti-viral cellular and humoral responses. It is necessary to find out if the failure of the antiviral CTLs to clear HIV-1 is caused by the viral proteins.

(iii) *The impact of HIV-1 infection on the human innate and adaptive immune responses.*

In dealing with the question of why allergic people are not allowed to be vaccinated with smallpox vaccine (vaccinia virus), Becker [29] reviewed the studies on the effects of aeroallergens presentation by LCs/DCs in patients of atopic dermatitis. It was reported that. T helper 2 (Th2) cells are activated by allergens to release cytokines IL-4, IL-5, and IL-13. IL-4 directs B cells to synthesize anti-allergen IgE molecules and inhibits the ability of T helper 1 cells to produce IL-2, IL-12, and IFN-γ that stimulate CTL activity. It is important to know that the synthesis of IgE in HIV-1 infected people, according to Wright et al. [43], markedly increases during infection indicating a bad prognosis. It was also reported by Secord [44] that children infected with HIV-1 that are nonprogressors responded to infection with the synthesis of anti-HIV-1 IgE antibodies, which reacted with the HIV-1 structural proteins gp120, p24, and p17, but not with known environmental allergens.

In 1993, Clerici and Shearer [45] published a hypothesis that the reason for the synthesis of IgE in HIV-1 infected people is due to a change in the activities of Th1 cells and Th2 cells. The authors hypothesized that a “Th1?Th2 switch” occurs after HIV-1 infection causing the inhibition of Th1 cell activity to synthesize IFN-γ and IL-12 and a decline in the activity of CTLs. The cytokine synthesize by Th2 cells is probably activated by the viral proteins.

Comparison of the “Th1?Th2 Switch” in HIV-1 patients to the immune disfunction in atopic allergy patient reveals a marked similarity between the two disease conditions. It led

Becker [46] to suggest that the HIV-1 protein have allergen-like amino acid domains that when presented by dendritic cells preferentially activate the Th2 cells to induce and release IL-4, IL-5, and IL-13 cytokines. The Th2 cytokine IL-4 inhibits the activity of Th1 cells and cause the inhibition of CTL induction [45].

The HIV-1 strains that start the infection in humans are slow replicating viruses that use the CCR5 chemokine co-receptor for entry into $CD4^+$ T cells. These viruses cannot cause syncytia of infected cells that are designated as nonsyncytium inducing virus(NSI). However, IL-4 inhibits the synthesis of the CCR5 co-receptor molecules and induces the synthesis of CXCR4 co-receptor molecules, which select syncytium-inducing (SI) virus that is capable of rapid replication in $CD4^+$ T cells.

(iv) *The urgent need to save HIV-1 infected people form AIDS-related death.*

After twenty years of intensive research and more than 40,000 publications on HIV-1/AIDS, it is necessary to find treatments that will prevent the virus-induced immune deficiency that causes the AIDS patient to lose hope.

Based on the above, Becker [46] hypothesized that the approach to curing HIV-1/AIDS should be to use drugs that inhibit the Th2 cell activity and activates the Th1 cytokine synthesis (Th2?Th1 reversion hypothesis [46]) that will revive the antiviral cellular immune response. Drugs that are immune response modifiers may have the desired effect.

Studies on the immune response modifier imiquimod (designated S-26308, R-837, which was developed and is manufactured by 3M Pharmaceuticals, St. Paul, MN 55144, USA) reported that the drug induces antiviral activity against HSV-1, cytomegalovirus in infected humans, and arbovirus infections in rodents [47]. Imiquimod modulates the innate immune response by activating plasmacytoid dendritic cells (DCs) via binding to Toll-like receptor 7 (TLR7) inducing the synthesis of interferon α (IFNα) and the proinflammatory Th1 cytokines that activate the cell-mediated immunity. In addition, imiquimod inhibited Th2 cell synthesis of IL-4 and IL-5 [48].

In regard to protective HIV-1 vaccines, it is logical to assume that viral proteins that will lack the allergen-like domains may be better immunogens that will induce antiviral CTLs and neutralizing antibodies.

4. Conclusions

a. The beginning of the twenty-first century has been marred by international terrorism and the emergence of a natural human epidemic caused by a new coronavirus. In addition, the HIV-1/AIDS epidemic that started in the 1980s continues to spread around the world without delay. The terror activities cause anxiety that perhaps biological weapons like smallpox (variola major) virus will be used against human populations. If such a bio-terrorist act should materialize, the world population may be subjected to an epidemic that will require an international effort under the guidance of the WHO to stop the spread of the pathogenic bio-terror agent.

There is a need for the scientific societies and universities to demand that scientists and science students will become aware of their responsibility in preventing the misuse of biological sciences. An international "Science for Peace" movement must be organized to convince UNESCO to start a program to influence universities around the world to plan courses that will train young scientists to know their ethical responsibilities in their future scientific work. In addition, the following Science for Peace Oath, as suggested in the Jerusalem Statement on Science for Peace (1997), should be accepted at all universities and taken by all scientists upon receiving their degrees: "My scientific activities will be for the benefit of humanity." If these ideas are implemented, the lives of future generations will be safer than ours.

It goes without saying that the Biological Weapon Convention should be finalized and signed.

b. We have learned from past epidemics that the international efforts of research scientists, health authorities, and health services are essential to combat microbial epidemics. The international research activities to isolate and characterize the causative SARS pathogen, monitoring of SARS patients, protection of the medical staff in hospitals from infection by the SARS virus, as well as the monitoring of travelers in airports worldwide for signs of infection and the coordination of these activities by the WHO prevented virus spread and the epidemic that started during November 2002 was over six months later.

c. An effective intervention program to stop the spread of smallpox was achieved by countries in which smallpox was endemic and travelers carried the virus to all continents. With special funds for this program, volunteers used the smallpox vaccine to immunize the entire population in endemic areas in India and the Horn of Africa. However, the current voluntary smallpox vaccination program in the United States to protect large populations against bioterror attack with smallpox brought up a major difficulty, namely, the large number of people suffering from allergies, immune deficiencies that are prohibited from vaccination. To overcome this problem, the search for a cure for this allergy must become a priority.

d. All what was learned from the successful control of virus pandemics (influenza, poliomyelitis, smallpox, measles, and SARS) had failed in the attempts to stop the HIV-1/AIDS pandemic. Treatments of HIV-1 infected people with mixtures of antiviral drugs have prolonged life spans of the infected individuals, but efforts to develop anti-HIV vaccines have failed.

Becker's hypothesis on the need to achieve a "Th2?Th1 Reversion" that inhibits Th2 activity in HIV-1/AIDS patients by the use of immune response modifiers that revive the antiviral T helper 1 cell activity may provide a new approach to save HIV-1 infected people form AIDS.

References

[1] Becker Y., Vary J., Ramirez J.M., and participants of the Second International Symposium on Science for Peace, *The Jerusalem Statement on Science for Peace, the International Symposia on Science for Peace*, in Volume 3, Science for Peace Series, Editors: Becker Y., Kouzminov V., and assisted by Santesso R., published by UNESCO-Venice Office, pp. 13–16, 1997.

[2] *Science for the twenty-first century — a new commitment: declaration on science and the use of scientific knowledge*, Science Agenda-Framework for Action. Published by UNESCO, Paris, 2000.

[3] World Health Organization, *The global eradication of smallpox: final report of the global commission for the certification of smallpox eradication*, Geneva, Switzerland: World Health Organization, 1980.

[4] Fauci A.S., "Smallpox vaccination policy — the need for dialogue," *N Eng J Med* 346, 1319–1320, 2002.

[5] Toth T., *The biological weapons convention: a historical perspective in the possible use of biological weapons by terrorist groups: scientific legal and international implication*, published by the International Centre for Genetic Engineering and Biotechnology, Triest, Italy, pp. 9–11, 2002.

[6] Alibek K. and Handelman S., *Biohazard*, Random House Australia Limited, 1999.

[7] Enserink M., "Avian flu outbreak sets off alarm bells," *Science* 300, 718, 2003.

[8] "Sars from the kitchens of Guandung to the Metropol Hotel (Hong Kong) to the rest of the world," *Haaretz* 28/4/2003, p. A4 [Hebrew].

[9] Rota A., Oberste M.S., Monroe S.S., Nix W.A., Campagnoli R., Icenogle J.P., et al., www.science.express.org/1May2003/Page1/10.1126/Science1085952.

[10] Marra M.A., Jones S.J.M., Astell C.R., Holt R.A., Brooks-Wilson A., Butterfield Y.S.N., et al., "The genome sequence of SARS-associated corona virus," www.scienceexpress.org/1May2003/Page1/10.1126/Science.1085943.
[11] Ksiazek T.G. et al., "A novel coronavirus associated with severe acute respiratory syndrome," *New E J Med* 348, 1953–1966, 2003.
[12] Holmes K.V. and Enjuanes L., "the SARS corona virus: a postgenomic era, *Science* 300, 1377–1378, 2003.
[13] Ensernik M., *Clues to the animal origin of SARS*?
[14] Dye C., and Gay N., "Modeling the SARS epidemic," www.scienceexpress.org/23May2003/Page1/10.1126/science.1086925.
[15] Anand K., Zielbuhm J., Wadhulani P., Mester J., Hilgenfeld R., "Corona virus main proteinase ($3Cl^{pro}$) structure: basis for design of anti-SARS," www.sciencepress.org/13May2003/Page1/10.1126/science1085658.
[16] Hopkins D.R., "Ramses V: earliest known victim?," *World Health* May 1980, p. 220.
[17] Elgood C.A., *Medical history of Persia and the Eastern Caliphate*, The University Press Cambridge, 1951.
[18] Acuna-Soto R., Stahle D.W., Cleaveland M.K., and Therrell M.D., "Megadraught and megadeath in 16th century Mexico," *Energ Infect Dis* 8, 360–362, 2002.
[19] Joklik W.K., *Principles of Animal Virology*, Appelton-Century-Crafts, New York, p. 229, 1980.
[20] Henderson D.A., "A victory for all mankind," *World Health*, May 1980, pp. 3–5.
[21] Mahler H., "Towards the year 2000, *World Health*, May 1980, p. 19.
[22] Alibek K. with Handelman S., *Biohazard*, Random House Australia (Pty) Limited, p. 111, 1999.
[23] Mair M. and Borio L., "Key information regarding smallpox vaccine," *Biosecurity and Bioterrorism* 1, 53–54, 2003.
[24] Fox J.L., "Understanding poxviruses still a major challenge," *ASM News* 68, 589, 2002.
[25] Trinidade G. de S., da Fouseca F.G., Marquis J.T., Noguera M.L., et al., "Arasatuba virus: a vaccinelike virus associated with infection in humans and cattle, *Emerg Infect Dis* 9, 155–160, 2003.
[26] Engler R.J.M., Kenner J., Leung D.Y.M., "Smallpox vaccination: risk considerations for patients with atopic dermatitis," *J Allergy Clin Immunol* 110, 357–365, 2002.
[27] Moss B. "Poxvirus Virulence," *ASM News* 69, 107–108, 2003.
[28] Sharma D.P., Ramsay A.J., Maguire D.J., Rolf M.S., and Ramshaw I.A., "Interleukin-4 mediates down regulation of antiviral cytokine expression and cytotoxic T-lymphocyte responses and exacerbates vaccinia virus infection in-vitro," *J Virol* 70, 7103–7107, 1996.
[29] Becker Y., "Vaccinia virus pathogenicity in atopic dermatitis is caused by allergen induced T helper 2 immune response that prevents the antiviral cellular and humoral immunity, *Virus Genes* 27(1), In Press.
[30] Fauci A.S., "Smallpox vaccination policy — the need for dialogue, *N Eng J Med* 346, 1319–1320, 2002.
[31] Essex M., and Kanki P.J., "the Origins of the AIDS virus," *Scientific American*, pp. 44–51, October 1988.
[32] Gao F. et al., "Origin of HIV-1 in the chimpanzee Pan troglodytes," *Nature* 397, 436–441, 1999.
[33] Marx P.A., Li Y., Lerche N.W., Sutjipto S., Gettie A., Yee J.A., Brotman B.H., Prince A.M., Hanson A., Webster R.G., et al., "Isolation of a simian immunodeficiency virus related to human immunodeficiency virus type 2 from a west African pet sooty mangabey," *J. Virol* 65, 4480–4485, 1991.
[34] Barre-Sinoussi F., Cherman J.-C., Rey F., Nugeyere M.T., Chamarit S., Gruest J., Dauguet C., Axler-Blir C., Brun-Vezinet F., Rouzioux C., Rosenbaum W., and Montagnier L., "Isolation of a T-lymphotropic retrovirus from a patient at risk for acquired immune deficiency syndrome (AIDS)," *Science* 220, 868–871, 1983.
[35] Stover J. et al., "Can we reverse the HIV/AIDS pandemic with an expanded response?," *Lancet* 360, 73–77, 2002.
[36] Schwartländer B., Garnett E., Walker N., and Anderson R., "AIDS in the new millenium," *Science* 289, 64–67, 2000.
[37] Najera R., Delgado E., Pérez-Alvarez L., and Thomson M.M., "Genetic recombination and its role in the development of the HIV pandemic," *AIDS* 16 (Suppl. 4) S3–S16, 2002.
[38] Sabin A.B., "Improbability of effective vaccination against human immunodeficiency virus because of its intracellular transmission and rectal portal of entry," *Natl Acad Sci USA* 89, 8852–8855, 1992.
[39] Chinen J. and Shearer W.T., "Molecular virology and immunology of HIV infection," *J Allergy Clinical Immunol* 110, 189–198, 2002.
[40] Cohen J., "Vaccine results lose significance under scrutiny," *Science* 299, 1495, 2003.
[41] Cohen J., "Disappointing data scuttle plans for large scale AIDS vaccine trial," *Science* 295, 1616–1617, 2002.

[42] Baba T.W., Liska V., Khimami A.H., Ray N.B., Dailey P.J., Pennick D., Ruprect R.M., "Live attenuated, multiply deleted simicin immunodeficiency virus causes AIDS in infant and adult macques," *Nature Med* 5, 194–203, 1999.

[43] Wright D.N., Nelson R.P., Ledford D.K., Fernandez-Caldas E., Trudeau W.L., and Lockey R.F., "Serum IgE and human immunodeficiency virus (HIV) infection," *J Allergy Clin Immunol* 85, 445–452, 1990.

[44] Secord E.A., Kleiner G.I., Auci D.L., Smith-Norowitz T., Chice S., Finkelstein A., Nowakowski M., Fikring S., and Durkin H.G., "IgE against HIV proteins in clinically healthy children with HIV disease," *J Allergy Clin Immunol* 98, 979–984, 1996.

[45] Clerici M. and Shearer G.M., "A Th1?Th2 switch is a critical step in the etiology of HIV infection," *Immunol Today* 14, 107–111, 1993.

[46] Becker Y. "The changes in the T helper 1 (Th1) and T helper 2 (Th2) cytokine balance during HIV-1 infection are indicative of an allergic response to viral proteins and may be reversed by Th2 cytokine inhibitors and immune response modifiers," *Virus Genes* 28(1), in press 2004.

[47] Wagner T.L., Horton V.L., Carlson G.L., Myhre P.E., Gibson S.J., Imberson L.M., and Tomai M.H., "Induction of cytokines in cynomolgus monkeys by immune response modifiers imiquimod, S27609 and S-28463," *Cytokine* 9, 837–845, 1997.

[48] Stanley M.A., "Imiquimod and imidazoquinolones: mechanism of action and therapeutic potential," *Clin Exper Dermatol* 27, 571–577, 2002.

Emerging Biological Threat
G. Berencsi et al. (Eds.)
IOS Press, 2005

Contributors

1) S.K. Alkhovsky,
D.I. Ivanovsky Institute of Virology RAMS, Gamaleya Str. 16. 123098 Moscow, Russia,
Tel.: +7 (095) 190-2872, Fax: +7 (095) 190-2867

2) V.A. Aristova,
D.I. Ivanovsky Institute of Virology RAMS, Gamaleya Str. 16. 123098 Moscow, Russia,
Tel.: +7 (095) 190-2872, Fax: +7 (095) 190-2867

3) Abdu F. Azad,
University of Maryland School of Medicine,
Baltimore, MD 21201

4) Dunja Z. Babič,
Institute of Microbiology and Immunology, Medical Faculty,
Zaloška 4, 1000 Ljubljana, Slovenia

5) Professor Yechiel Becker, Director,
UNESCO-HUJ International School of Molecular Biology and Microbiology, The Institute of Microbiology, the Department of Molecular Virology, Faculty of Medicine, The Hebrew University of Jerusalem, POB 12272, Jerusalem 91120, Israel,
Tel.: 972-2-6758394, Fax: 972-2-6784010

6) György Berencsi, MD, PhD,
Division of Virology,
"B. Johan" National Center for Epidemiology, Gyali Str. 2–6. H-1097 Budapest, Hungary

7) Joachim J. Bugert,
Department of Medical Microbiology, University of Wales College of Medicine,
Heath Park, Cardiff, CF14 4XN, UK

8) E.I. Burtseva,
D.I. Ivanovsky Research Institute of Virology Russian Academy of Medical Science
Gamaleya Str. 16, 123098 Moscow, Russia

9) Erik Busch-Petersen,
Baxter Vaccine AG, Industriestrasse 67, A-1220 Vienna, Austria

10) B.Ts. Bushkieva,
State Center of Sanitary-Epidemiological Inspection, Elista, Russia

11) A.M. Butenko,
D.I. Ivanovsky Institute of Virology RAMS, Gamaleya Str. 16. 123098 Moscow, Russia, Tel.: +7 (095) 190-2872, Fax: +7 (095) 190-2867

12) Saulius Caplinskas, MD, PhD, D.Sc.,
Lithuanian AIDS Center, Vilnius, Lithuania

13) Alíz Czeglédi,
"B. Johan" National Center for Epidemiology, Division of Virology,
Budapest, Hungary

14) Vito G. DelVecchio,
Institute of Molecular Biology and Medicine, University of Scranton, Scranton, Pa. 18510, and Vital Probes, Inc. 1300 Old Plank Road, Mayfield Pa, 18433

15) P.G. Deryabin,
D.I. Ivanovsky Institute of Virology RAMS, Gamaleya Str. 16. 123098 Moscow, Russia, Tel.: +7 (095) 190-2872, Fax: +7 (095) 190-2867 and
I.M. Sechenov Moscow Medical Academy, Moscow, Russia

16) Beverly K. Dyas,
Laboratory of Molecular Immunology,
Army Medical Research Institute of Infectious Diseases, Frederick, MD USA

17) A.F. Dzharkenov,
State Center of Sanitary-Epidemiological Inspection, Astrakhan, Russia

18) Richard E. Eberle,
Dept. Veterinary Pathobiology, College of Veterinary Medicine, Oklahoma State University

19) Matt Ewert, MS, CT (ASCP),
Center for Biological Defense, USF

20) Gábor Faludi,
Medical Research Institute of the Hungarian Defence Forces,
Budapest, Hungary

21) Emőke Ferenczi,
"B. Johan" National Center for Epidemiology, Division of Virology,
Budapest, Hungary

22) Zdenek Filip,
Federal Environmental Agency, Langen Building, Paul Ehrlich Strasse 29, 63225 Langen, Present address: Hegstrauch 7, 35463 Fernwald, Germany

23) I.V. Galkina,
D.I. Ivanovsky Institute of Virology RAMS, Gamaleya Str. 16. 123098 Moscow, Russia, Tel.: +7 (095) 190-2872, Fax: +7 (095) 190-2867

24) M. Garstang,
Acambis plc, Peterhouse Technology Park, Cambridge, UK

25) V.L. Gromashevsky,
D.I. Ivanovsky Institute of Virology RAMS, Gamaleya Str. 16. 123098 Moscow, Russia, Tel.: +7 (095) 190-2872, Fax: +7 (095) 190-2867

26) D.J. Gubler,
Division of Vector-Borne Infection Diseases, National Center for Infectious Diseases, Centers for Disease Control and Prevention, Fort Collins, Colorado, USA

27) Péter Gyarmati,
Division of Virology,
"B. Johan" National Center for Epidemiology, Gyáli str. 2–6, Budapest, Hungary

28) Susanne Herrmann,
Federal Environmental Agency, Langen Building, Paul Ehrlich Strasse 29, 63225 Langen, Germany

29) Akbar S. Khan,
Molecular Engineering Team, US Army Soldier and Biological Chemical Command,
5183 Blackhawk Road, Edgewood CB Center, APG, MD 21010-5424.
Present address: DTRA-CBM, 8725 John J. Kingman Road, MS 6201, Fort Belvoir
VA 22060-6201

30) Zoltán Kis,
Division of Virology,
"B. Johan" National Center for Epidemiology, Gyáli str. 2–6, Budapest, Hungary

31) R. Kinney,
Division of Vector-Borne Infection Diseases, National Center for Infectious Diseases, Centers for Disease Control and Prevention, Fort Collins, Colorado, USA

32) Teri Kissner,
Laboratory of Molecular Immunology,
Army Medical Research Institute of Infectious Diseases, Frederick, MD USA

33) Boštjan J. Kocjan,
Institute of Microbiology and Immunology, Medical Faculty,
Zaloška 4, 1000 Ljubljana, Slovenia

34) A.I. Kovtunov,
State Center of Sanitary-Epidemiological Inspection, Astrakhan, Russia

35) Jaromir Kubat,
Research Institute for Crop Production, Dept. of Soil Biology, Drnovska 507, 162 00 Prague 6, Czech Republic

36) L.N. Kulikova,
State Center of Sanitary-Epidemiological Inspection, Astrakhan, Russia
Andras Lakos, MD, PhD
Centre for Tick-borne Diseases, Budapest, Hungary

37) George V. Ludwig,
US Army Medical Research Institute of Infectious Diseases, Fort Detrick, Maryland, USA

38) Dimitri K. Lvov, MD, PhD, D.Sc., *General Director,*
I. Ivanovsky Institute of Virology RAMS, Gamaleya Str. 16. 123098 Moscow, Russia
Tel.: +7 (095) 190-2872, Fax: +7 (095) 190-2867 and
I.M. Sechenov Moscow Medical Academy, Moscow, Russia

39) Maria M. Medveczky,
Dept. Medical Microbiology and Immunology, University of South Florida, Tampa, FL USA

40) Peter G. Medveczky, MD, PhD
Dept. Medical Microbiology and Immunology, University of South Florida, Tampa, FL USA

41) Jeff Morgan,
Applied Ordnance Technology, Inc., 25 Center Street, Stafford, VA 22554

42) T.M. Moskvina,
D.I. Ivanovsky Institute of Virology RAMS, Gamaleya Str. 16. 123098 Moscow, Russia,
Tel.: +7 (095) 190-2872, Fax: +7 (095) 190-2867

43) J. Pastorek,
Institute of Virology, Slovak Academy of Sciences,
84505 Bratislava, Slovak Republic

44) Mario Poljak,
Institute of Microbiology and Immunology, Medical Faculty,
Zaloška 4, 1000 Ljubljana, Slovenia

45) Ljudmilla Priimägi, MD, PhD, D.Sc.,
National Institute for Health Development, Virology Department, Tallinn, Estonia

46) A.G. Prilipov,
D.I. Ivanovsky Institute of Virology RAMS, Gamaleya Str. 16. 123098 Moscow, Russia,
Tel.: +7 (095) 190-2872, Fax: +7 (095) 190-2867

47) Gábor Rácz,
University of New Mexico, Department of Biology,
Albuquerque, NM

48) Julius Rajčáni,
Institute of Virology, Slovak Academy of Sciences,
84505 Bratislava, Slovak Republic

49) Col. László Rókusz MD,
Central Military Hospital of Hungarian Defense Forces, Budapest, Hungary

50) G.K. Sadykova,
D.I. Ivanovsky Institute of Virology RAMS, Gamaleya Str. 16. 123098 Moscow, Russia, Tel.: +7 (095) 190-2872, Fax: +7 (095) 190-2867

51) Kamal U. Saikh,
Laboratory of Molecular Immunology,
Army Medical Research Institute of Infectious Diseases, Frederick, MD USA

52) Younes Ali Saleh,
Division of Virology,
"B. Johan" National Center for Epidemiology, Gyáli str. 2–6, H-1097 Budapest, Hungary; On leave from the Faculty of Medicine, University of Garyounis, Benghazi, Libyan Jamahiriya

53) E.I. Samokhvalov,
D.I. Ivanovsky Institute of Virology RAMS, Gamaleya Str. 16. 123098 Moscow, Russia, Tel.: +7 (095) 190-2872, Fax: +7 (095) 190-2867

54) H.M. Savage,
Division of Vector-Borne Infection Diseases, National Center for Infectious Diseases, Centers for Disease Control and Prevention, Fort Collins, Colorado, USA

55) Katja Seme,
Institute of Microbiology and Immunology, Medical Faculty,
Zaloška 4, 1000 Ljubljana, Slovenia

56) A.G. Shatalov,
D.I. Ivanovsky Institute of Virology RAMS, Gamaleya Str. 16. 123098 Moscow, Russia, Tel.: +7 (095) 190-2872, Fax: +7 (095) 190-2867

57) M.Yu. Shchelkanov,
D.I. Ivanovsky Institute of Virology RAMS, Gamaleya Str. 16. 123098 Moscow, Russia, Tel.: +7 (095) 190-2872, Fax: +7 (095) 190-2867

58) A.N. Slepushkin,
D.I. Ivanovsky Research Institute of Virology Russian Academy of Medical Science
Gamaleya Str. 16. 123098 Moscow, Russia

59) Peter J. Stopa,
US Army Edgewood Chemical Biological Center, 5183 Blackhawk Road
AMSRD-ECB-ENP-MC, E3549, Aberdeen Proving Ground, MD 21010-5424 USA

60) Afroz Sultana,
Laboratory of Molecular Immunology,
Army Medical Research Institute of Infectious Diseases, Frederick, MD USA

61) Wieslaw Swietnicki,
Laboratory of Molecular Immunology,
Army Medical Research Institute of Infectious Diseases, Frederick, MD USA

62) Tatjana Tallo,
National Institute for Health Development, Virology Department
Tallinn, Estonia

63) Valentina Tefanova,
National Institute for Health Development, Virology Department
Tallinn, Estonia

64) Robert G. Ulrich,
Laboratory of Molecular Immunology,
Army Medical Research Institute of Infectious Diseases, Frederick, MD USA

65) V.E. Usachev,
D.I. Ivanovsky Institute of Virology RAMS, Gamaleya Str. 16. 123098 Moscow, Russia,
Tel.: +7 (095) 190-2872, Fax: +7 (095) 190-2867

66) James Valdes,
Molecular Engineering Team, US Army Soldier and Biological Chemical Command,
5183 Blackhawk Road, Edgewood CB Center, APG, MD 21010-5424

67) L.N. Vlassova,
D.I. Ivanovsky Research Institute of Virology Russian Academy of Medical Science
Gamaleya Str. 16. 123098 Moscow, Russia

68) A.G. Voronina,
D.I. Ivanovsky Institute of Virology RAMS, Gamaleya Str. 16. 123098 Moscow, Russia,
Tel.: +7 (095) 190-2872, Fax: +7 (095) 190-2867

69) George E. Wright,
GLSynthesis, Worcester, MA USA

70) K.B. Yashkulov,
State Center of Sanitary-Epidemiological Inspection, Elista, Russia

71) L.V. Zlobina,
State Center of Sanitary-Epidemiological Inspection, Astrakhan, Russia

Author Index

Alkhovsky, S.K. 33
Aristova, V.A. 33
Azad, A.F. 147
Babič, D.Z. 77
Becker, Y. 159
Berencsi, G. 3, 43, 50
Bugert, J.J. 146
Burtseva, E.I. 26
Busch-Petersen, E. 157
Bushkieva, B.Ts. 33
Butenko, A.M. 33
Caplinskas, S. 10
Czeglédi, A. 43
DelVecchio, V.G. 109
Deryabin, P.G. 33
Dyas, B.K. 93
Dzharkenov, A.F. 33
Eberle, R.E. 58
Ewert, M. 101
Faludi, G. 43, 117
Ferenczi, E. 43
Filip, Z. 148
Galkina, I.V. 33
Garstang, M. 157
Gromashevsky, V.L. 33
Gubler, D.J. 33
Gyarmati, P. 3, 50
Herrmann, S. 148
Ivanova, V.T. 26
Jankovics, I. 117
Khan, A.S. 104
Kinney, R. 33
Kis, Z. 3, 50
Kissner, T. 93
Kocjan, B.J. 77
Kopcsó, C.I. v
Kovtunov, A.I. 33
Kubat, J. 148
Kulikova, L.N. 33
Lakos, A. 60
Ludwig, G.V. 84
Lvov, D.K. 33
Lvov, D.N. 33
Medveczky, M.M. 58
Medveczky, P.G. 58
Mezey, I. 43
Morgan, J. 125
Moskvina, T.M. 33
Pastorek, J. 62
Poljak, M. 77
Priimägi, L. 20
Prilipov, A.G. 33
Rácz, G. 43
Rajčáni, J. 62
Rókusz, L. 136
Sadykova, G.K. 33
Saikh, K.U. 93
Saleh, Y.A. 50
Samokhvalov, E.I. 33
Sarkadi, J. 117
Savage, H.M. 33
Seme, K. 77
Shatalov, A.G. 33
Shchelkanov, M.Yu. 33
Slepushkin, A.N. 26
Stopa, P.J. 125
Sultana, A. 93
Swietnicki, W. 93
Tallo, T. 20
Tefanova, V. 20
Ulrich, R.G. 93
Usachev, V.E. 33
Valdes, J. 104
Visontai, I. 117
Vlassova, L.N. 26
Voronina, A.G. 33
Wright, G.E. 58
Yashkulov, K.B. 33
Zelenka, G. 117
Zlobina, L.V. 33